Mathematics for Industry

Volume 28

Aims & Scope

The meaning of "Mathematics for Industry" (sometimes abbreviated as MI or MfI) is different from that of "Mathematics in Industry" (or of "Industrial Mathematics"). The latter is restrictive: it tends to be identified with the actual mathematics that specifically arises in the daily management and operation of manufacturing. The former, however, denotes a new research field in mathematics that may serve as a foundation for creating future technologies. This concept was born from the integration and reorganization of pure and applied mathematics in the present day into a fluid and versatile form capable of stimulating awareness of the importance of mathematics in industry, as well as responding to the needs of industrial technologies. The history of this integration and reorganization indicates that this basic idea will someday find increasing utility. Mathematics can be a key technology in modern society.

The series aims to promote this trend by (1) providing comprehensive content on applications of mathematics, especially to industry technologies via various types of scientific research, (2) introducing basic, useful, necessary and crucial knowledge for several applications through concrete subjects, and (3) introducing new research results and developments for applications of mathematics in the real world. These points may provide the basis for opening a new mathematics oriented technological world and even new research fields of mathematics.

More information about this series at http://www.springer.com/series/13254

Robert S. Anderssen · Philip Broadbridge
Yasuhide Fukumoto · Kenji Kajiwara
Matthew Simpson · Ian Turner
Editors

Agriculture as a Metaphor for Creativity in All Human Endeavors

Springer

Editors
Robert S. Anderssen
Mathematics, Information and Statistics
CSIRO
Acton, ACT
Australia

Philip Broadbridge
La Trobe University
Melbourne, VIC
Australia

Yasuhide Fukumoto
Kyushu University
Fukuoka
Japan

Kenji Kajiwara
Kyushu University
Fukuoka
Japan

Matthew Simpson
Queensland University of Technology
Brisbane, QLD
Australia

Ian Turner
Queensland University of Technology
Brisbane, QLD
Australia

ISSN 2198-350X ISSN 2198-3518 (electronic)
Mathematics for Industry
ISBN 978-981-10-7810-1 ISBN 978-981-10-7811-8 (eBook)
https://doi.org/10.1007/978-981-10-7811-8

Library of Congress Control Number: 2017963860

Printed on acid-free paper

This Springer imprint is published by Springer Nature
The registered company is Springer Nature Singapore Pte Ltd.
The registered company address is: 152 Beach Road, #21-01/04 Gateway East, Singapore 189721, Singapore

Preface

This book is the proceedings of the conference "Forum Math-for-Industry 2016 (FMfI2016)" held at Queensland University of Technology (QUT), for November 21–23, 2016, for which the unifying theme was "Agriculture as a metaphor for creativity in all human endeavors" and collects together selected papers presented there.

The agricultural process of planting a seed, fertilizing, growing, and harvesting has a clear parallel with the application of mathematics to a practical problem. The seed becomes the question being asked, the fertilization is the conceptualization of the mathematical framework within which to seek the answer, the growing is the solution process, and the harvesting is the articulation and implementation of the answer.

In agriculture, the breeding of the seed to plant involves genetics; the germination of the plant involves moisture alone; the growth involves the interaction between the biology and environment with soil, water, and weather the key drivers; the survival depends on its ability to respond to viral and fungal infections and stress challenges; and the flowering and setting of the seed for the next generation depend on the occurrence of environmental cues.

For understanding the processes and mechanisms involved with each of these steps, mathematical modeling is central. This is reflected in the emergence of new mathematically focused agriculture endeavors such as "precision agriculture," "smart agriculture analytics," and "digital agriculture."

The success of agriculture practice relies fundamentally on its interconnections with and dependence on biology and the environment. Both play fundamental roles including the adaption of biology to cope with environmental challenges of biotic and abiotic stresses and global warming. FMfI2016 explored the contribution of mathematics within the framework of the interaction of agriculture with biology and the environment.

The contents of this volume report on productive and successful interaction between industry and mathematicians, as well as on the cross-fertilization and collaboration that occurred. The book contains excellent examples of the roles of mathematics in innovation and, thereby, the importance and relevance of the concept Mathematics_FOR_Industry.

We would like to thank the participants of the forum and the members of the Scientific Board of the Forum, especially Troy Farrell, Matthew Simpson, and Ian Turner of QUT. Without their cooperation and support, we would never have experienced the great excitement and success of the forum. Moreover, we would like to express our deep appreciation for the great help of the conference secretaries during the preparation and organization of the forum, and Chiemi Furutani for the proceedings.

Fukuoka, Japan
April 2017

Yasuhide Fukumoto
On behalf of the Organizing Committee of the Forum Math-for-Industry 2016 and the Editorial Committee of the Proceedings

FMI2008	FMI2009	FMI2010	FMI2011	FMI2012	FMI2013	FMfI2014	FMfI2015	FMfI2016
Tokyo	Fukuoka	Fukuoka	Honolulu	Fukuoka	Fukuoka	Fukuoka	Fukuoka	Brisbane
Sep.16-17	Nov.9-13	Oct.21-23	Oct.24-28	Oct.22-26	Nov.4-8	Oct.27-31	Oct.26-30	Nov.21-23
The 1st Forum: Consortium Math For Industry	Casimir Force, Casimir Operators and the Riemann Hypothesis	Information Security, Visualization, and Inverse Problems, on the basis of Optimization Techniques	TSUNAMI - Mathematical Modelling- Using Mathematics for Natural Disaster Prediction, Recovery and Provision for the Future	Information Recovery and Discovery	The Impact of Applications on Mathematics	Applications + Practical Conceptualization + Mathematics = fruitful Innovation	The Role and Importance of Mathematics in Innovation	Agriculture as a metaphor for creativity in all human endeavors

Asia Pacific Consortium of Mathematics for Industry

Forum "Math-for-Industry" 2016

Agriculture as a metaphor for creativity in all human endeavors

November 21-23, 2016

Science and Engineering Centre, Gardens Point Campus, Queensland University of Technology, Brisbane, Australia

INVITED SPEAKERS

Alona Ben-Tal	(Massey University, New Zealand)
Peter Caley	(CSIRO, Australia)
Adelle Coster	(The University of New South Wales, Australia)
Bronwyn Harch	(Queensland University of Technology, Australia)
Kei Hirose	(Kyushu University, Japan)
Tony Jakeman	(Australian National University, Australia)
Julien Lerat	(Bureau of Meteorology, Australia)
Osamu Maruyama	(Kyushu University, Japan)
James McCaw	(The University of Melbourne, Australia)
Kerrie Mengersen	(Queensland University of Technology, Australia)
Mikio Murata	(Tokyo University of Agriculture and Technology, Japan)
Toshiyuki Nakagaki	(Hokkaido University, Japan)
Mark Nelson	(University of Wollongong, Australia)
Takashi Okayasu	(Kyushu University, Japan)
Tristan Perez	(Queensland University of Technology, Australia)
Julia Piantadosi	(University of South Australia, Australia)
Mick Roberts	(Massey University, New Zealand)
Akiko Satake	(Kyushu University, Japan)
Jason Sharples	(The University of New South Wales, Australia)
Jo Simpson	(Fonterra, New Zealand)
Takeshi Tsuji	(Kyushu University, Japan)
Hayato Waki	(Kyushu University, Japan)
Peter Waterhouse	(Queensland University of Technology, Australia)
Yoshihiro Yamanishi	(Kyushu University, Japan)
Shintaro Akamine	(Kyushu University, Japan)
Elliot Carr	(Queensland University of Technology, Australia)
Anton Gulley	(The University of Auckland, New Zealand)

ORGANIZING COMMITTEE

Bob Anderssen	(CSIRO, Australia)
Philip Broadbridge	(La Trobe University, Australia)
Troy Farrell	(Queensland University of Technology, Australia) Chair
Yasuhide Fukumoto	(IMI, Kyushu University, Japan)
Alexandra Hogan	(Australian National University, Australia)
Kenji Kajiwara	(IMI, Kyushu University, Japan)
Robert McKibbin	(Massey University, New Zealand)
Matthew Simpson	(Queensland University of Technology, Australia)
Ian Turner	(Queensland University of Technology, Australia)

Supported by

Asia Pacific Consortium of Mathematics for Industry
QUT School of Mathematical Sciences, Science and Engineering Faculty
Institute of Mathematics for Industry, Kyushu University

FMfI2016 Organizing Office

QUT School of Mathematical Sciences, Science and Engineering Faculty
Gardens Point Campus fmfi2016@imi.kyushu-u.ac.jp
http://apcmfi.org/fmfi2016/index.html Tel: +61-7-3138-2308

Forum "Math-for-Industry" 2016
Agriculture as a metaphor for creativity in all human endeavors
Science and Engineering Centre, Garden's Point Campus, Queensland University of Technology, Brisbane, Australia

Time	20-Nov SUNDAY	21-Nov MONDAY	22-Nov TUESDAY	23-Nov WEDNESDAY
9:00-9:30		Tristan Perez Queensland University of Technology, Australia	Mark Nelson University of Wollongong, Australia	Alona Ben-Tal Massey University, New Zealand
9:30-10:00		Adelle Coster The University of New South Wales, Australia	Julien Lerat Bureau of Meteorology, Australia	Hayato Waki Kyushu University, Japan
10:00-10:30		Bronwyn Harch Queensland University of Technology, Australia	Takashi Okayasu Kyushu University, Japan	Takeshi Tsuji Kyushu University, Japan
10:30-11:00		Morning Tea	Morning Tea	Morning Tea
11:30-12:00		James McCaw The University of Melbourne, Australia	Jason Sharples The University of New South Wales, Australia	POSTER PRESENTATION
12:00-12:30		Mick Roberts Massey University, New Zealand	Julia Piantadosi University of South Australia, Australia	
12:30-13:00		Mikio Murata Tokyo University of Agriculture and Technology, Japan	Peter Caley CSIRO, Australia	Lunch
13:00-13:30		Lunch	Lunch	
13:30-14:00				Yoshihiro Yamanishi Kyushu University, Japan
14:00-14:30		Kerrie Mengersen Queensland University of Technology, Australia	EXCURSION	Peter Waterhouse Queensland University of Technology, Australia
14:30-15:00		Kei Hirose Kyushu University, Japan		Akiko Satake Kyushu University, Japan
15:00-15:30		Anthony Jakeman Australian National University, Australia		Afternoon Tea
15:30-16:00		Afternoon Tea		Joanne Simpson Fonterra, New Zealand
16:00-16:30		Shintaro Akamine Kyushu University, Japan		Osamu Maruyama Kyushu University, Japan
16:30-17:00		Elliot Carr Queensland University of Technology, Australia		Toshiyuki Nakagaki Hokkaido University, Japan
17:00-17:30	APCMfI Meeting	Anton Gulley The University of Auckland, New Zealand		
17:30-18:00		IMI-IAB Meeting		
18:00-18:30				WORKSHOP DINNER
18:30-19:00	FMfI Executive Meeting			
19:00-19:30				
19:30-20:00				
20:00-				

Biology	Agriculture	Environment	Early Career Researchers	Other Topics

Contents

The Shape of Things to Come—Using Geometric and Morphometric Analyses to Identify Archaeological Starch Grains

Adelle C. F. Coster and Judith H. Field

Abstract Starch grains are tell-tale characteristics of plants that can remain long after the decomposition of the rest of the material. The understanding of historical plant use, for sustenance and plant-based medicines, as well as agricultural practices is enhanced by the identification of residual starch remains. Classifications, however, have previously relied on expert identification using largely subjective features. This can be enormously time consuming and subject to bias. A method has been developed to construct robust classifiers for starch grains of unknown origin based on their geometrical and morphometric features. It was established to allow insight into plant food use from archaeological remains but could be used in many different contexts.

Keywords Mathematics-for-Industry · Starch grains · Identification
Geometric analysis · Morphometric analysis

1 Introduction

Starch grains can be preserved for millennia on grinding tools and surfaces. They are insoluble granules of carbohydrates that build up in plants. They have an initial growth point, or hilum, and then layers of material build up around this point. As they are simply carbohydrate, it is not possible to tell from their chemical composition the species from which they came. However, due to the different cellular structures of different plants, the starch grains have shapes that are characteristic of the particular species and organelle from which they are derived. The features of the starch grains from known plant species can form a reference library. The within-species

A. C. F. Coster (✉)
School of Mathematics and Statistics, UNSW, Sydney, Australia
e-mail: A.Coster@unsw.edu.au

J. H. Field
School of Biological, Earth and Environmental Sciences, UNSW, Sydney, Australia
e-mail: judith.field@unsw.edu.au

R. S. Anderssen et al. (eds.), *Agriculture as a Metaphor for Creativity in All Human Endeavors*, Mathematics for Industry 28,
https://doi.org/10.1007/978-981-10-7811-8_1

features and variability can then be utilised to both describe the species and also as a discriminator to classify unknown samples, recovered from artefacts, soils or other materials.

2 Experimental Methods

The starch grains can be recovered from the artefact, soil or other material, suspended in a mixture of glycerol and water and imaged using a brightfield, differential interference contrast microscope. In the current approach, we extract the two-dimensional maximum-projection-area grain shape (identified as the in-focus edge of the grain), or region of interest, ROI, from the light micrograph. Additionally, the hilum point is identified. In our method, a hybrid approach to edge detection is employed which combines automatic outline analysis with some expert intervention to finalise the outline and hilum positioning. This is because purely automated edge detection is difficult across a large variety of images—the grains have different depths of fields, and the assessment of the in-focus edge can be unclear when the morphology causes different shadowings [1]. An accurate edge is important as the ROI is used to obtain the discriminative features to classify the grains. Morphological dilation and erosion, common image analysis techniques for edge detection and object separation, degrade the features of the ROI.

2.1 Geometric and Morphometric Features

The ROI and hilum location are used to calculate the geometric and morphometric features of the grain. These include the area, perimeter and centre of mass of each grain. Hilum offset measures encode the position of the centre of mass and then compared to the hilum position. The maximum length through the hilum of the grain has previously been shown to be a good discriminant for some plant taxa [2]. This maximum length line running through the hilum also provides a reference angle from which other features can be observed. Other characteristics such as circularity and other shape matching measures and curvature metrics can also be calculated.

In our case, the digitised ROI has very closely spaced edge points. For completeness, however, we approximate the periphery radius as a piecewise linear function of the angle about the hilum position relative to the maximum length line.

The starch grains in our studies have no convexity issues, so it is possible to expand the periphery radius as a radial Fourier series. Thus, we can generate a model from a truncated sum that can approximate the grain shape. In practice, we have found that the perimeter features can be represented by as few as five terms (see [6]). The radial harmonic components of the Fourier decomposition are characteristic of the grain shape and can be used to discriminate between different species.

However, if convexity is an issue, other decompositions can be employed such as wavelet shape (e.g. [3]), multi-scale fractal dimension or curvature scale-space analysis (e.g. [4]).

3 Starch Grain Classification

In order to discriminate between species, however, it is important to have a comprehensive, well-curated reference library with which to compare grains of unknown origin. The reference set for the classifier also needs to be appropriate to the geographic region [5]. The reference grains need to be sourced from the appropriate plant parts—e.g. seeds, fruiting bodies and tubers. It is also good practice to use multiple samples for each species, to, for instance, account for variation in environmental conditions and their possible impact on the size of the grains. The reference set for the classifier should also be, for our archaeological purposes, of important economic plants from the appropriate location, altitude and climate from the people occupying the site at the period of interest. Evidence of use by the people using the land is also important.

Samples of starch grains from the reference species were analysed to calculate the predictor variable values for each grain in the population. Within species, the shapes will have some variance and may also vary with orientation. We have found that the within-species variation appears to be captured by approximately 100 grains in the 80 species we have analysed so far, [6, 7], ensuring a statistically significant result and allowing the decomposition of the species into sub-grouping as required.

The morphometric measures of the two-dimensional projections are used to determine the classifiers that were best able to discriminate between the grains. Classifiers were considered for various choices of the predictor variables, the classifier type, the training set (the species to be considered) and the output classes—the species or their sub-groupings, and in latter case, the method of resulting species prediction.

Series of classifiers were built taking different combinations of the predictor variable, which included the maximum length, area, perimeter, circularity, hilum offset measures and the Fourier components. Other measures such as the shape matching variables were investigated but found to be of lesser discriminative value than those listed above.

Possible combinations of predictor variables can be explored by calculating a MANOVA of sets of predictor variables for the reference species (or sub-species). Separation of the species by the MANOVA is an indicator that it may be feasible to discriminate between the species in a classifier. It may be possible, however, to still positively identify the presence of a subset of the training set, even if others are indistinguishable.

In these investigations, the classifiers were broadly discriminant, nearest neighbour and decision tree; however, other algorithms including neural networks and support vector machines could also be used. In deciding which algorithms to use, there is a trade-off in performance and the number of design parameters that need to be explored.

The method to assess the classifiers depends on the type of output desired. For instance, if a distribution of a particular species is to be estimated, then the classifier needs to be designed to best classify all the unknown grains. Designing a classifier to obtain a very high confidence, true-positive identification for some of the unknown grains may be to the detriment of certain classifications for others.

It is the latter option which was taken in assessing the archaeologically provenanced starch grains. Whilst the distribution of species was of interest, it was deemed to be of more importance to know that there was evidence for a particular species being present in the samples—this would, for instance, indicate that a particular plant was consumed if the grain was found on a grinding stone or in tooth calculus.

The accuracy of the different classifiers constructed was assessed both by re-substitution and cross-validation. Re-substitution assesses the accuracy of a classifier that has been constructed using all the data. Each (known) grain in the training set is identified using the classifier and the rate of true and false positive classifications determined. Given the natural within-species variation and possible overlaps between species, it is unlikely that 100% accuracy will be achieved, even when re-substituting the grains used in the classifier construction. Re-substitution accuracy, however, does not necessarily correspond to the accuracy of the classifier when presented with a grain that was not used in its construction. This accuracy can be assessed by cross-validation, whereby the training set is randomly partitioned into two subsets. One is used to construct the classifier, and the another is withheld, and then used to test the classifier performance. Note that the species of origin for all the grains is known. The process of partitioning, training and testing is repeated multiple times to validate the cross-fold error in classifying the withheld subset. As the process withholds part of the training data each iteration, the classifiers constructed are (a) not identical to that constructed with all the training data and (b) may mean that the cross-validation error may be an overestimate. In general, this means that the accuracy of the cross-validation classifiers is less than that obtained using re-substitution. Both measures are, however, useful in indicating the classifier performance and can be used in concert to determine the which of the suite of classifiers is of most utility for a given problem.

In developing the methodology for starch grain identification, we further developed the idea of the cross-fold validation for this system. Given we are interested in true-positive, confident predictions we consider a reverse cross-fold validation error. Rather than the usual cross-validation, where a number of grains of a known species are classified and the number of mis-classifications measured, here we look at classification results from all the grains tested. For a given species classification, we then determine how many were not actually that species. The classifier can perform differently for different species. We developed, [7], a measure to encode the dependence of the true-positive rate on the classifier score for the different output species called the positive prediction value, PPV. The PPV can then be used to choose the classifier and determine the confidence of the results. If the prediction score corresponds to, say, $PPV = 0.9$ for that species, then 90% of the predictions with scores above this level were correctly classified. Lesser scores, corresponding to lower PPV values, mean that the unknown could possibly be the predicted species, albeit with less confidence.

Once the different classifiers are constructed for the training set, using different algorithms and combinations of predictor variables, and assessment can be made of their performance, discarding those with low re-substitution success and high cross-fold validation errors. Further analysis of the performance for the individual species within the training set via the PPV and the individual species cross-validation errors then allows us to choose the "best" classifier for the problem at hand.

As part of our studies, once the classifier was optimised and the unknown grains classified, the unknowns were furthermore re-analysed by an expert microscopist, skilled in starch analysis. The quantitative system outlined above does not take categorical or subjective attributes into account when performing the classifications. Subjective features of starch grains include the presence of lamellae and pitting of the grain surface. Taking the cross-fold validation confidence and PPV for the predicted species into account, each prediction was deemed to be validated, probable or a false positive. We found a high correspondence with the predicted values, except in some species of similar geometry where subjective features such as lamellae were prominent in some species and not others.

4 Discussion

The methodology has been used to create starch grain classifiers for a number of important archaeological and ethnographic studies. We have identified that the inhabitants of North Queensland rainforests undertook complex processing of some starchy nuts, which were otherwise toxic [7]. Trading patterns and plant use in the highlands of Papua New Guinea have been identified from residue remains on grinding tools. The foraging and consumption of seeds have been investigated from ground stone artefacts from Woomera in South Australia [8].

Whilst the development of the collection of reference grains has been a time-consuming venture, as we are maintaining user input into the collection of the individual ROIs, this library then becomes accessible to multiple studies. If the plant species is present or is an appropriate inclusion in the training set for a given study, these can be deployed immediately. As an evolving collection, this represents a valuable curated resource. It then remains to analyse the newly sources unknown grains—once they have been found in the field!

The methodology is currently being employed to identify starch grains recovered from a variety of archaeological contexts in Australia, Papua New Guinea, and the USA. It is not, however, limited to archaeological samples, and could, for instance, be used to study the provenance of produce such as honey if starch grains in the sample can be identified.

The cultivation of wheat has been integral to human civilisation for millennia. In modern times, different cultivars of wheat produce 'soft' and 'hard' flours, suitable for different food preparations. It is planned to characterise the wheat grains producing the different flour qualities and correlate the starch grain features with the kernel strength and hardness profiles [9, 10].

The approach taken here could also be used beyond starch, at, for instance, larger physical scales. Grain shape is a key factor affecting the mechanical properties of granular materials. It has long been of interest in sedimentology (see for instance [11–14]). An accurate classifier of grain shape could be used to quickly and accurately provide information on the contents of seed samples such as wheat, oats, rye and barley. This could be coupled with grain handling operations to recover and sort mixed grains or grade samples. Some work has been done in this area [15] and could be extended to use the approaches of the method developed for starch identification.

References

1. J. Wilson, K. Hardy, R. Allen, L. Copeland, R. Wrangham, M. Collins, Automated classification of starch granules using supervised pattern recognition of morphological properties. J. Archaeol. Sci. **37**, 594–604 (2010)
2. B. Lance, J. Field, Intra-taxonomic variability in starch reference collections and the implications for ancient starch studies. *Unpublished Manuscript on file in School of BEES* (UNSW Australia, 2006), pp. 21
3. H. Drolon, F. Druaux, A. Faure, Particles shape analysis and classification using the wavelet transform. Pattern Recogn. Lett. **21**, 473–482 (2000)
4. João B. Florindo, André R. Backes, and Odemir M. Bruno, Leaves shape classification using curvature and fractal dimension image and signal processing, ed. by A. Elmoataz et al. in *Proceedings of the 4th International Conference, ICISP 2010*, Quêbec, Canada, June 30–July 2, 2010. LNCS 6134 (2010), pp. 456462
5. J. Field, Reference collections for starch studies, in *Ancient Starch Analysis*, ed. by R. Torrence, H. Barton (Left Coast Press, California, 2006), pp. 95–113
6. A.C.F. Coster, J.H. Field, What starch grain is that?—A geometric morphometric approach to determining plant species origin. J. Archaeol. Sci. **58**, 9–25 (2015)
7. J.H. Field, L. Kealhofer, R. Cosgrove, A.C.F. Coster, Human-Environment Dynamics during the Holocene in the Australian Wet Tropics of NE Queensland: a starch and phytolith study. J. Anthropol. Archaeol. **44**, 216–234 (2016)
8. S. Luu, B. Stephenson, A.C.F. Coster, P.T.T. Nguyen, J. Field, Residue analysis of ground stone artefacts from Woomera, South Australia: a pilot study. A Report Prepared for Mr Tim Owen, GML Heritage and the Kokatha Aboriginal Community, Woomera, SA, 30 Sept 2015
9. R.S. Anderssen, R. Haraszi, Characterizing and exploiting the rheology of wheat hardness. Eur. Food Res. Technol. **229**, 159–174 (2009)
10. R. Haraszi, A. Juhasz, M. Sissons, M. Rakszegi, L. Tamas, R.S. Anderssen, Rheological hardness index for assessing hardness of hexaploids and durums. Cereal Chem. **90**, 430–438 (2013)
11. W.C. Krumbein, F.J. Pettijohn, *Manual of Sedimentary Petrography* (Appleton-Century-Crofts, New York, 1938)
12. J.C. Griffiths, *Scientific Method in Analysis of Sediments* (McGraw-Hill Inc., New York, 1967)
13. G. Lees, A new method for determining the angularity of particles. Sedimentology **3**, 2–21 (1964)
14. R. Ehrlich, B. Weinberg, An exact method for characterization of grain shape. J. Sediment. Petrol. **40**, 205–212 (1970)
15. J. Paliwal, N.S. Visen, D.S. Jayas, N.D.G. White, Comparison of a neural network and a non-parametric classifier for grain kernel identification. Biosyst. Eng. **85**(4), 405–413 (2003)

New Mathematical Models of Antimalarial Drug Action to Improve Drug Dosing Regimens

James M. McCaw, Pengxing Cao, Sophie Zaloumis and Julie A. Simpson

Abstract *Plasmodium falciparum* malaria remains a major threat to global public health. Artemisinin-based combination therapies—a critical component of current control strategies—are at risk of failure due to the emergence of artemisinin resistance. To extend the life of artemisinin-based therapies, it is crucial that we develop a better understanding of how they act to reduce parasitemia in the host. Recent laboratory-based experiments have demonstrated that parasites respond to the cumulative, rather than instantaneous, drug concentration. This observation directly challenges the standard paradigm of pharmacokinetic–pharmacodynamic (PK–PD) modelling. Here, we introduce a generalisation to the PK–PD model which accounts for cumulative exposure. Parasites accumulate 'stress', which translates into an effective killing rate which can vary with both drug concentration and exposure time. Our model indicates how drug-resistant parasites may avoid killing. Through simulation, we explore alternative drug dosing strategies that may overcome drug resistance.

Keywords Mathematics for Industry · Biological modelling · Malaria
Antimalarial drugs

1 Introduction

Artemisinin derivatives (ART) provide the first-line treatment for *falciparum* malaria, a major parasitic disease affecting millions of people every year [1]. Their extensive use over the past decade has dramatically reduced the burden of malaria on human populations. However, over recent years, clinical signs of drug resistance

J. M. McCaw (✉) · P. Cao
School of Mathematics and Statistics, The University of Melbourne,
Melbourne, VIC 3010, Australia
e-mail: jamesm@unimelb.edu.au

S. Zaloumis · J. A. Simpson
Melbourne School of Population and Global Health, The University of Melbourne,
Melbourne, VIC 3010, Australia

R. S. Anderssen et al. (eds.), *Agriculture as a Metaphor for Creativity in All Human Endeavors*, Mathematics for Industry 28,
https://doi.org/10.1007/978-981-10-7811-8_2

have become established in South-East Asia and ART therapy is now at risk of failure [2].

Pharmacokinetic–pharmacodynamic (PK–PD) models—which combine models of blood antimalarial drug concentrations with models of parasite replication dynamics—have been used extensively to study the mechanisms of action of drugs, interpret clinical trial data on alternative dosing regimens and guide the development of drug dosing policy [3].

Laboratory experiments, conducted by collaborators at the Bio21 Institute (Melbourne), have established that the dynamics of drug killing are complex [4, 5]. Not only do parasites display stage-specific sensitivity to ART, the rate of parasite killing appears to depend upon the cumulative exposure of parasites to drug, rather than the instantaneous drug concentration. Furthermore, in experiments with 'drug-resistant' parasite strains, a clear loss of sensitivity to the drug was observed during the ring stage of the parasite's life cycle.

In this context, understanding the mechanisms and dynamics of drug-induced parasite killing requires the development of new PK–PD models of drug activity and parasite response [3]. The development of improved models may prove crucial in optimising drug regimens to either overcome or delay the onset of drug resistance and improve clinical outcomes.

Here, we introduce a model of parasite killing in the presence of a time-varying drug concentration and extend the PK–PD modelling paradigm to account for drug accumulation effects.

2 Model

During blood-stage *Plasmodium falciparum* infection, the parasites go through a 48-h asexual reproductive life cycle. We consider the number of parasitised red blood cells, $N(a, t)$, of age a at time t to evolve according to

$$\frac{\partial N(a,t)}{\partial t} + \frac{\partial N(a,t)}{\partial a} = -kN(a,t) \,, \tag{1}$$

where k is the (drug-induced) parasite killing rate, which in general will be a function of the detailed history of drug exposure. The domain of a is the time from formation of an infected red blood cell to its time of rupture a_r (usually 48 h). We have a boundary condition for asexual reproduction $N(0, t) = rN(a_r, t)$ ($r \approx 10$), indicating that parasites released from a single ruptured red blood cell infect (on average) 10 susceptible red blood cells.

From this general formulation (which will be used to perform simulations of in vivo parasite dynamics at the end of this paper), we simplify to consider a tightly age-synchronised population of parasites of age $\bar{a}$ as was used in the in vitro experiments of Klonis [4] and Dogovski [5]. Drug responses are (empirically) observed to be well described by Hill function kinetics:

$$k(C) = \frac{k_{\max} C^{\gamma}}{K_c^{\gamma} + C^{\gamma}} .$$

To capture the effects of cumulative drug exposure, we model the maximal killing rate, $k_{\max}$ and half-maximal concentration K_c to be functions of an accumulated parasite 'stress', S, which increases as follows whenever drug, $C(t)$, is present at a concentration higher than some (small) critical value, C^*:

$$\frac{dS}{dt} = \lambda(1 - S) .$$

The stress, S, is immediately reset to zero when $C(t)$ drops back below C^*.

With $k_{\max} = \alpha S$ and $K_c = \beta_1(1 - S) + \beta_2$, with α, β_1 and β_2 positive constants, and assuming the presence of drug at a concentration $C(t) > C^*$ for the entire experimental assay, we obtain

$$k_{\max} = \alpha \left(1 - e^{-\lambda t}\right)$$
$$K_c = \beta_1 e^{-\lambda t} + \beta_2 ,$$

and so the number of parasites $\bar{N}(t)$ (of initial age $\bar{a}$) surviving at time t is given by:

$$\bar{N}(t) = \bar{N}_0 \exp\left[-\int_0^t k(C(\tau), S(\tau))\, d\tau\right] .$$

The in vitro experiments [4, 5] exposed parasites to drug pulses of a particular duration (T_d) and particular (initial) concentration C_0. The half-life of in vitro drug decay was also measured. Rather than measure counts of parasites directly, the experiments provide a relative measure of parasite survival based on the number that survive until rupture, producing 'offspring' (with expansion factor r (see above)) in the next generation. This measure is called the *viability* and is constrained to lie in [0, 1]. After some manipulation (and with details of the experimental procedure [4]), it can be shown that the viability is given by

$$V(C_0, T_d) = \exp\left[-\int_0^{T_d} k(C(\tau), S(\tau))\, d\tau\right] . \tag{2}$$

2.1 Results

We used model (2) to fit the available age/stage-specific in vitro data. For each of the four parasite stages (early rings, mid-rings, early trophozoites and late trophozoites), we obtained (stage-specific) estimates for λ, α, β_1 and β_2. With these estimates—which show strong evidence for drug accumulation effects (i.e. λ is small, data not

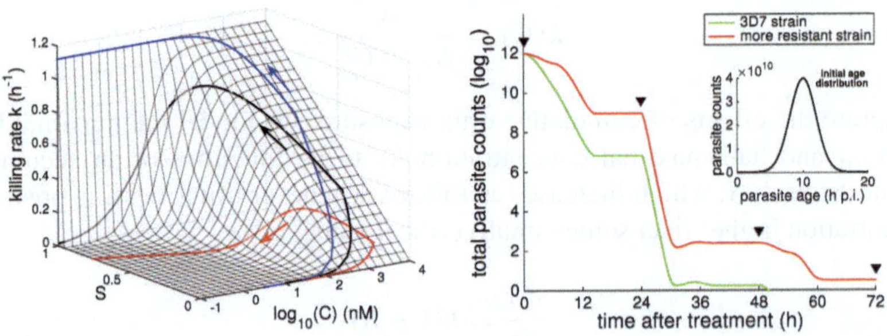

Fig. 1 **a** The killing rate surface as a function of drug concentration C and accumulated stress S and a projection of the trajectory for the effective killing rate on that surface for three values of the stress accumulation rate λ (details in text). **b** Simulation of the parasite load (from simulation of the full model (1)) under a standard 24 hourly treatment regimen for two strains: drug-sensitive 3D7 ($\lambda = 0.37$) and a hypothetical drug-resistant strain ($\lambda = 0.1$)

shown)—we then simulated realistic in vivo PK–PD curves and explore the effects of drug accumulation.

Figure 1a shows how the overall killing rate k evolves for the mid-ring stage of the parasite life cycle given a typical PK (drug concentration) time profile. The trajectory in black is for $\lambda = 0.37$, as estimated from the in vitro data. Curves for slower accumulation of 'stress' ($\lambda = 0.1$, red) and more rapid accumulation ($\lambda = 1.0$, blue) are also presented to highlight the potential biological importance of drug accumulation. It has been suggested that drug resistance may manifest as an increased tolerance to drug for the parasite [5]. We model this possibility as a reduction in λ for the mid-ring stage of the life cycle. Figure 1b illustrates the dramatic effect on the parasite load over time resulting from this increased drug tolerance for a realistic scenario of multiple drug doses (following the standard drug regimen as recommended by the World Health Organisation).

Having established how an increased tolerance to stress manifests as a delay in clearance, we now use the model to explore alternative drug dosing regimens. Figure 2 presents results for two widely suggested alternative regimens: increasing dose concentration and increasing dose frequency. We apply these alternative dosing regimens to a simulated 3D7 infection (i.e. $\lambda = 0.37$). For an initial parasite distribution in the host that is primarily rings (a) or trophozoites (b), we observe a clear benefit in twice daily dosing. In contrast, marginal benefit is obtained through increased dose concentration (4mg/kg vs. 2mg/kg).

2.1.1 Conclusions

Based on detailed in vitro experiments and an extension to the traditional PK–PD modelling framework, we have explored the potential role for drug accumulation effects in antimalarial activity. Our findings provide new insight into the mechanisms

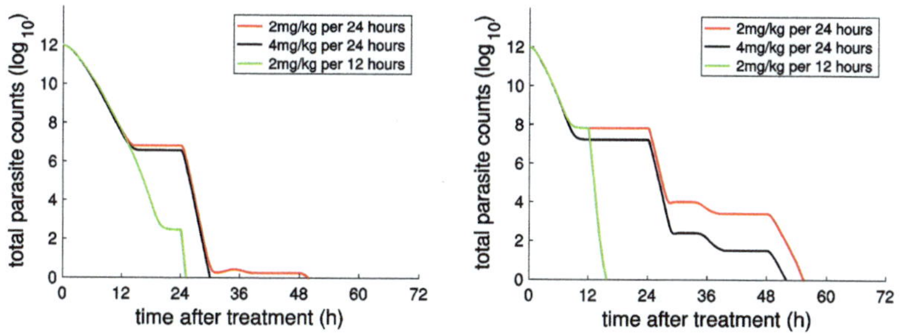

Fig. 2 A comparison of alternative drug regimens. **a** Parasite counts over time for the baseline (2mg/kg per 24 h; red), increased dose concentration (4mg/kg; black) and increased dose frequency (2mg/kg per 12 h; green). Increasing the dose has a minimal effect on parasite count as the drug concentration is sufficiently high in the baseline scenario. However, increasing the dosing frequency to twice daily has a dramatic and positive impact, shortening the time to resolution of infection by between 12 and 24 h. **b** As in (**a**) but for an older initial parasite distribution (primarily trophozoites). The improvement obtained by twice daily dosing is even more dramatic

of drug-induced parasite killing and an enhanced predictive platform for evaluating the likely efficacy of alternative ART dosing regimens.

Acknowledgements We thank Leann Tilley and her team (Bio21, The University of Melbourne) for access to data. Pengxing Cao and Sophie Zaloumis were supported by National Health and Medical Research Council project and Centre for Research Excellence funding.

References

1. R.T. Eastman, Fidock Da, Artemisinin-based combination therapies: a vital tool in efforts to eliminate malaria. Nat. Rev. Microbiol. **7**, 864–874 (2009)
2. A.M. Dondorp, R.M. Fairhurst, L. Slutsker, J.R. Macarthur, J.G. Breman et al., The threat of artemisinin-resistant malaria. N. Engl. J. Med. **365**, 1073–1075 (2011)
3. J.A. Simpson, S. Zaloumis, A.M. DeLivera, R.N. Price, J.M. McCaw, Making the most of clinical data: reviewing the role of pharmacokinetic-pharmacodynamic models of anti-malarial drugs. AAPS J (2014)
4. N. Klonis, S.C. Xie, J.M. McCaw, M.P. Crespo-Ortiz, S.G. Zaloumis et al., Altered temporal response of malaria parasites determines differential sensitivity to artemisinin. Proc. Natl. Acad. Sci. U.S.A. **110**, 5157–5162 (2013)
5. C. Dogovski, S.C. Xie, G. Burgio, J. Bridgford, S. Mok et al., Targeting the cell stress response of Plasmodium falciparum to overcome artemisinin resistance. PLoS Biol. **13**, e1002132 (2015)

Biodiversity and the Ecology of Emerging Infectious Diseases

M. G. Roberts and J. A. P. Heesterbeek

Abstract The question of how biodiversity influences the emergence of infectious diseases is the subject of ongoing research. A set of nonlinear differential equations is been used to explore the interactions between ecology and epidemiology. The model allows for frequency-dependent transmission of infection within host species, and density-dependent transmission between species, via the environment or a vector. Three examples are discussed. It is shown that removing a pathogen may increase a consumer population, decreasing its resource. It is then shown that the presence of a pathogen could enable a predator and a prey species to coexist. Finally the dilution effect, by which increasing biodiversity reduces the transmission of an infectious disease, is investigated.

Keywords Biodiversity · Ecology · Epidemiology · Infectious diseases

1 Introduction

Emerging infection diseases present a major threat to world health. On average, two new species of human virus are reported each year [1], most having an animal origin [2–5]. Recent examples are SARS [6], swine flu [7] and avian influenza [8, 9]. In 2014, Ebola virus re-emerged from a bat reservoir [10, 11], causing a major epidemic [12–15]. Climate change could lead to *Aedes* mosquitoes establishing in New Zealand [16, 17], and with them dengue and Zika viruses [18].

M. G. Roberts (✉)
Institute of Natural and Mathematical Sciences, New Zealand Institute for Advanced Study, Infectious Disease Research Centre, Massey University, Private Bag 102 904 NSMC, 0745 Albany, Auckland, New Zealand
e-mail: m.g.roberts@massey.ac.nz

J. A. P. Heesterbeek
Faculty of Veterinary Medicine, Department of Farm Animal Health, University of Utrecht, Yalelaan 7, 3584 CL Utrecht, The Netherlands
e-mail: j.a.p.heesterbeek@uu.nl

R. S. Anderssen et al. (eds.), *Agriculture as a Metaphor for Creativity in All Human Endeavors*, Mathematics for Industry 28,
https://doi.org/10.1007/978-981-10-7811-8_3

Large complex ecosystems interacting at random are almost certain to be unstable [19]. Adding structure to the community matrix produces a different picture [20–23], competitive interactions are stabilising, whereas mutualism is destabilising [20]. A major component of an ecosystem is the food web: the network of feeding interactions among species. Adding pathogens increases the web's complexity [24–28], and parasites have been described as the *dominant* or *missing* links [29, 30]. An infection may make prey easier to catch, or unpalatable to a predator, or reduce a predator's hunting ability, but the overall influence of pathogens on an ecosystem may be unexpected [31]. The influence of ecosystem dynamics on epidemiology can also be unexpected [32] and may lead to a pathogen *jumping* host species causing a pandemic.

We present a model that describes how an infectious disease can modify the dynamics of host and non-host species, and how changes in ecosystem dynamics can modify the epidemiology of a pathogen. We illustrate our model with three examples. In the first, eliminating a pathogen led to an increase in biodiversity, whereas in the second the presence of a pathogen is necessary to maintain a prey–predator relationship. The third example directly addresses the dilution effect—how a change in biodiversity may result in a change in the dynamics of an infectious disease.

2 The Model

To model an infectious disease on a food web infected with a single pathogen of interest, we define N_i to be the abundance of species $i \in \Omega$, I_i/N_i to be the proportion of species i infected, and $S_i = N_i - I_i$ the abundance of susceptible hosts of species i. The equations for the population dynamics of the food web are

$$\begin{aligned}\frac{\mathrm{d}N_i}{\mathrm{d}t} &= \nu_i(N_i)N_i - \mu_i(N_i)N_i - \alpha_i I_i - \sum_{j\in\mathcal{N}_i} \phi_{ij}\left(S_i + n_{ij}I_i\right)\left(S_j + o_{ij}I_j\right)\\ &\quad - \sum_{k\in\mathcal{P}_i} \psi_{ik}\left(S_i + q_{ik}I_i\right)\left(S_k + p_{ik}I_k\right) + \sum_{\ell\in\mathcal{Q}_i} \pi_{\ell i}\psi_{\ell i}\left(S_i + p_{\ell i}I_i\right)\left(S_\ell + q_{\ell i}I_\ell\right),\end{aligned}$$

where species i is born at the rate ν_i and dies at the rate μ_i, both functions of N_i, with increased death rate due to infection α_i. Species i competes for resources with species j when $j \in \mathcal{N}_i$, is eaten by species k when $k \in \mathcal{P}_i$ and eats species ℓ when $\ell \in \mathcal{Q}_i$. The variables ϕ_{ij}, ψ_{ik} and $\pi_{\ell i}$ account for competition for resources between species i and j, consumption of species i by species k and the benefit to species ℓ of consuming species i, respectively. All of these interactions may be modified if one or other of the species is infected with the pathogen.

The dynamics of the pathogen are expressed by

$$\frac{\mathrm{d}I_i}{\mathrm{d}t} = \beta_i \frac{S_i I_i}{N_i} - \mu_i(N_i)I_i - \alpha_i I_i - I_i \sum_{j\in\mathscr{N}_i} \phi_{ij} n_{ij} \left(S_j + o_{ij} I_j\right)$$
$$- I_i \sum_{k\in\mathscr{P}_i} \psi_{ik} q_{ik} \left(S_k + p_{ik} I_k\right) + S_i \sum_{\ell\in\mathscr{Q}_i} \gamma_{\ell i} q_{\ell i} \psi_{\ell i} I_\ell + \kappa_i S_i W$$

The model allows for three modes of transmission: frequency-dependent intra-species transmission at rate β_i, density-dependent transmission via the environment or an infected vector $\left(W = \sum_{m\in\Omega} r_m I_m\right)$ and transmission from prey to predator while feeding.

There are usually multiple steady states. The structure of the Jacobian matrix at the infection-free steady state decouples criteria for **ecological stability** and **epidemiological stability** [33].

$$\mathbf{J} = \begin{pmatrix} \mathbf{C} & \mathbf{D} \\ \mathbf{0} & \mathbf{H} \end{pmatrix}$$

The steady state is ecologically stable if the maximum real part of the eigenvalues of the community matrix is negative, $s(\mathbf{C}) < 0$. The steady state is epidemiologically stable if $s(\mathbf{H}) < 0$. The matrix $\mathbf{H}$ determines the stability of an ecological equilibrium to invasion by an infectious disease in *chronological time*. It can be decomposed $\mathbf{H} = \mathbf{T} + \boldsymbol{\Sigma}$ where $\mathbf{T}$ is the transmission matrix for the pathogen and $\boldsymbol{\Sigma}$ is the transition matrix. The **next-generation matrix** is $\mathbf{K} = -\mathbf{T}\boldsymbol{\Sigma}^{-1}$, and the basic reproduction number $\mathscr{R}_0$ is the spectral radius of $\mathbf{K}$ [34, 35]. If $\mathscr{R}_0 > 1$, the pathogen can invade the food web, and hence $\mathbf{K}$ determines epidemiological stability of the ecosystem *in generation time*.

Example 1 **A resource–consumer–pathogen system**. For this simple example, we assume that the pathogen only infects the consumer. The equations are

$$\begin{aligned}
\frac{\mathrm{d}N_1}{\mathrm{d}t} &= \nu_1(N_1)N_1 - \mu_1 N_1 - \psi N_1 \left(S_2 + pI_2\right) \\
\frac{\mathrm{d}N_2}{\mathrm{d}t} &= \nu_2(N_2)N_2 + \pi\psi N_1 \left(S_2 + pI_2\right) - \mu_2 N_2 - \alpha I_2 \\
\frac{\mathrm{d}I_2}{\mathrm{d}t} &= \beta \frac{S_2 I_2}{N_2} - \left(\mu_2 + \alpha\right) I_2
\end{aligned}$$

The Jacobian matrix at any infection-free state $(N_1, N_2, 0)$ simplifies to

$$\mathbf{J} = \begin{pmatrix} N_1 \nu_1'(N_1) & -\phi N_1 & \psi(1-p)N_1 \\ \pi\psi N_2 & N_2 \nu_2'(N_2) & -\pi\psi(1-p)N_1 - \alpha \\ 0 & 0 & \beta - \mu_2 - \alpha \end{pmatrix}$$

The community matrix $\mathbf{C}$ is the leading 2×2 sub-matrix of $\mathbf{J}$. It has negative trace and positive determinant, and hence it is stable. The infection-free equilibrium is unstable if $\mathscr{R}_0 = \frac{\beta}{\mu_2+\alpha} > 1$. The steady states of the system are plotted as functions

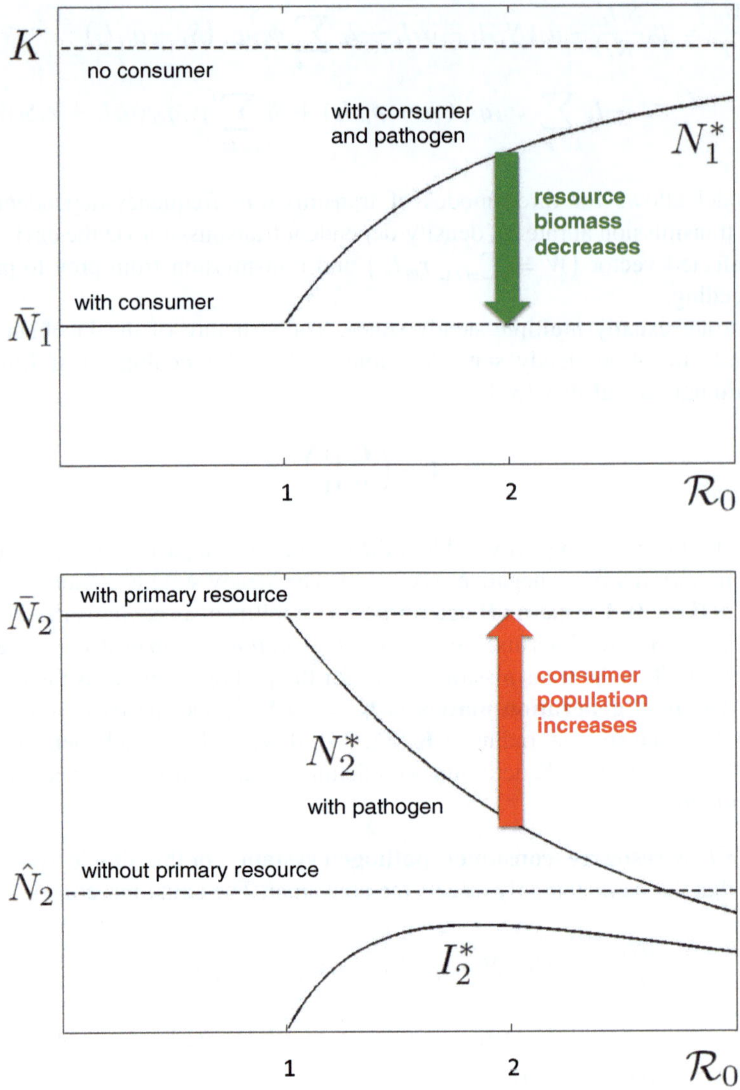

Fig. 1 Bifurcation diagram for the model presented in Example 1. Top: steady states for the resource plotted against the basic reproduction number $\mathscr{R}_0$. K is the steady state in the absence of the consumer, $\bar{N}_1$ in the presence of the consumer and N_1^* in the presence of consumer and pathogen. Bottom: steady states for the consumer plotted against $\mathscr{R}_0$. $\hat{N}_2$ is the steady state in the absence of the primary resource, $\bar{N}_2$ in the presence of the resource and N_2^* in the presence of resource and pathogen. I_2^* is the abundance of the pathogen. The effect of removing the pathogen is indicated by the arrows

of $\mathscr{R}_0$ in Fig. 1. Eliminating the pathogen from the consumer increases its abundance from N_2^* to $\bar{N}_2$. As a consequence, the resource biomass decreases from N_1^* to $\bar{N}_1$. This is consistent with observations in the Serengeti. Following the eradication of rinderpest, wildebeest numbers increased and the grass biomass decreased. As a consequence there were fewer fires, more trees, more giraffes and more predators [36]. These are further interactions that could have been included in a larger model.

Example 2 **A prey–predator–pathogen system**. For this example, the pathogen infects both prey (species 1) and predator (species 2), with transmission from predator to prey via environmental contamination. The model is

$$\begin{aligned}
\frac{\mathrm{d}N_1}{\mathrm{d}t} &= \nu_1(N_1)N_1 - \mu_1 N_1 - \alpha_1 I_1 - \psi\,(S_1 + qI_1)\,(S_2 + pI_2) \\
\frac{\mathrm{d}N_2}{\mathrm{d}t} &= \nu_2(N_2)N_2 + \pi\psi\,(S_1 + qI_1)\,(S_2 + pI_2) - \mu_2 N_2 - \alpha_2 I_2 \\
\frac{\mathrm{d}I_1}{\mathrm{d}t} &= \beta_1 \frac{S_1 I_1}{N_1} - (\mu_1 + \alpha_1)\,I_1 - \psi q I_1\,(S_2 + pI_2) + \kappa S_1\,(I_1 + rI_2) \\
\frac{\mathrm{d}I_2}{\mathrm{d}t} &= \gamma q \psi I_1 S_2 + \beta_2 \frac{S_2 I_2}{N_2} - (\mu_2 + \alpha_2)\,I_2
\end{aligned}$$

The basic reproduction number is the spectral radius of the next-generation matrix, $\mathscr{R}_0 = \rho\,(\mathbf{K})$, where

$$\mathbf{K} = \begin{pmatrix} \frac{\beta_1}{\mu_1+\alpha_1+\psi q\overline{N}_2} + \frac{\kappa\overline{N}_1}{\mu_1+\alpha_1} & \frac{\kappa\overline{N}_1}{\mu_2+\alpha_2} \\ \frac{\gamma q\psi\overline{N}_2}{\mu_1+\alpha_1+\psi q\overline{N}_2} & \frac{\beta_2}{\mu_2+\alpha_2} \end{pmatrix}$$

In the absence of prey–predator interaction, $\psi = 0$. The basic reproduction number in the prey is then $\mathscr{R}_0 = \frac{\beta_1+\kappa\widehat{N}_1}{\mu_1+\alpha_1}$, and in the predator $\mathscr{R}_0 = \frac{\beta_2}{\mu_2+\alpha_2}$. The possible steady states of the prey species are $\widehat{N}_1$ without predators; $\overline{N}_1$ with predators; N_1^* without predators with pathogen; and N_1^{**} with predators and pathogen. These are plotted as functions of the feeding rate ψ in Fig. 2. When the feeding rate of the predator is greater than a critical value ($\psi > \psi_{\mathrm{crit}}$), the prey species is driven to extinction unless the pathogen is present. Hence, the presence of the pathogen is necessary to keep the prey population viable.

Example 3 **The dilution effect**. Resolving the situations under which the dilution effect applies is an outstanding challenge in epidemiology [37]. The idea is that reducing biodiversity removes species that are hosts of a particular pathogen, hence increasing the risk of transmitting that pathogen to a new host, notably a human. The alternative is that removing hosts from an ecosystem reduces the viability of the pathogen, possibly driving it to extinction. A simple model with two prey species (1 and 2), two predator species (3 and 4) and one pathogen has been used to explore the dilution effect. The host population dynamics are described by

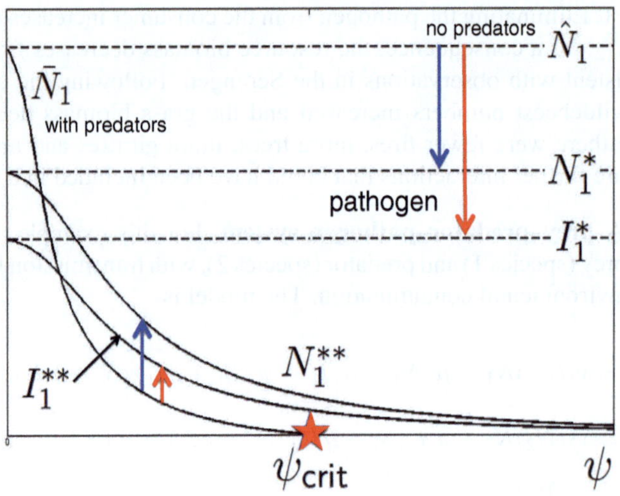

Fig. 2 Steady states of the prey population (species 1) in Example 2 plotted against the feeding rate of the predator ψ. The states are $\hat{N}_1$ without predators; $\overline{N}_1$ with predators; N_1^* without predators with pathogen; and N_1^{**} with predators and pathogen. The abundance of infected prey is I_1^* or I_1^{**} with predators. The effect of introducing the pathogen is indicated by the arrows

$$\left.\frac{\mathrm{d}N_i}{\mathrm{d}t}\right|_{i=1,2} = \nu_i N_i - \mu_i N_i - N_i \sum_{j=1,2} \phi_{ij} N_j - N_i \sum_{k=3,4} \psi_{ik} N_k$$

$$\left.\frac{\mathrm{d}N_i}{\mathrm{d}t}\right|_{i=3,4} = \nu_i N_i - \mu_i N_i - N_i \sum_{j=3,4} \phi_{ij} N_j + N_i \sum_{\ell=1,2} \pi_{\ell i} \psi_{\ell i} N_\ell$$

and the abundance of infected hosts by

$$\left.\frac{\mathrm{d}I_i}{\mathrm{d}t}\right|_{i=1,2} = \beta_i \frac{S_i I_i}{N_i} - \mu_i I_i - I_i \sum_{j=1,2} \phi_{ij} N_j - I_i \sum_{k=3,4} \psi_{ik} N_k + \kappa_i S_i W$$

$$\left.\frac{\mathrm{d}I_i}{\mathrm{d}t}\right|_{i=3,4} = \beta_i \frac{S_i I_i}{N_i} - \mu_i I_i - I_i \sum_{j=3,4} \phi_{ij} N_j + S_i \sum_{\ell=1,2} \gamma_{\ell i} \psi_{\ell i} I_\ell + \kappa_i S_i W$$

Preliminary results based on frequency-dependent transmission only show no dilution effect in prey species and an effect in predator species if increasing the prey population leads to an increased predator population. Adding transmission from prey to predator while feeding means the effects of population increase must exceed the effects of transmission to susceptibles through consuming prey. The results are more complicated with density-dependent transmission. We found a dilution effect in prey species in response to environmental dilution and a dilution effect in predator species under some restricted conditions. Exploring these effects is the subject of ongoing research.

3 Conclusion

A model described by a set of nonlinear differential equations has been used to explore the interactions between ecology and epidemiology. Three examples have been discussed. In the first, it was shown that removing a pathogen increased a consumer population and decreased the resource. The second example showed that the presence of a pathogen could enable a predator and prey species to coexist. Finally, the complex issue of the dilution effect was addressed. The question of how biodiversity influences the emergence of infectious diseases is the subject of ongoing research.

References

1. M.E.J. Woolhouse, H. Howey, E. Gaunt et al., Temporal trends in the discovery of human viruses. Proc. R. Soc. Sers. B **275**, 2111–2115 (2008)
2. N.D. Wolfe, C.P. Dunavan, J. Diamond, Origins of major human infectious diseases. Nature **447**, 279–283 (2007)
3. M.E.J. Woolhouse, How to make predictions about future infectious disease risks. Phil. Trans. R. Soc. Lond. B **366**, 2045–2054 (2011)
4. M.E.J. Woolhouse, D.T. Haydon, R. Antia, Emerging pathogens: the epidemiology and evolution of species jumps. Trends Ecol. Evol. **20**, 238–244 (2005)
5. M.E.J. Woolhouse, F. Scott, Z. Hudson et al., Human viruses: discovery and emergence. Phil. Trans. R. Soc. Lond. B **367**, 2864–2871 (2012)
6. A.R. McLean, R.M. May, J. Pattison et al., *SARS: A Case Study in Emerging Infections* (Oxford University Press, Oxford, 2005)
7. Y. Yang, J.D. Sugimoto, N.E. Basta et al., The transmissibility and control of pandemic influenza A (H1N1) virus. Science **326**, 729–733 (2009)
8. G. Milne, J. Kelso, H. Kelly, Strategies for mitigating an influenza pandemic with pre-pandemic H5N1 vaccines. J. R. Soc. Interface **7**, 573–586 (2010)
9. Y. Watanabe, M.S. Ibrahim, Y. Suzuki et al., The changing nature of avian influenza A virus (H5N1). Trends Microbiol. **20**, 11–20 (2012)
10. D.T.S. Hayman, Biannual birth pulses allow filoviruses to persist in bat populations. Proc. R. Soc. Sers. B **282**, 20142591 (2015)
11. R.K. Plowright, P. Eby, P.J. Hudson et al., Ecological dynamics of emerging bat virus spillover. Proc. R. Soc. Sers. B **282**, 20142124 (2014)
12. A. Camacho, A. Kucharski, S. Funk et al., Potential for large outbreaks of Ebola virus disease. Epidemics **9**, 70–78 (2014)
13. E.C. Hayden, The Ebola questions. Nature **514**, 554–557 (2014)
14. M.G. Roberts, Fast-spreading killers: how Ebola compares with other diseases. *The Conversation*. published online 10 Nov 2014
15. WHO Ebola Response Team, Ebola virus disease in West Africa the first 9 months of the epidemic and forward projections. N. Engl. J. Med. **371**, 1481–1495 (2014)
16. H. Bennett, R. Jones, G. Keating et al., Health and equity impacts of climate change in Aotearoa-New Zealand, and health gains from climate action. N. Z. Med. J. **127**, 16–31 (2014)
17. J.G. Derraik, D. Slaney, E.R. Nye et al., Chikungunya virus: a novel and potentially serious threat to New Zealand and the South Pacific islands. Am. J. Trop. Med. Hyg. **83**, 755–759 (2010)
18. D. Musso, V.M. Cao-Lormeau, D.J. Gubler, Zika virus: following the path of dengue and chikungunya? Lancet **386**, 243–244 (2015)

19. R.M. May, Will a large complex system be stable? Nature **238**, 413–414 (1972)
20. S. Allesina, S. Tang, Stability criteria for complex ecosystems. Nature **483**, 205–208 (2012)
21. I. Donohue, O.L. Petchey, J.M. Montoya et al., On the dimensionality of ecological stability. Ecol. Lett. **16**, 421–429 (2013)
22. A. Mougi, M. Kondoh, Diversity of interaction types and ecological community stability. Science **337**, 349–351 (2012)
23. A.-M. Neutel, J.A.P. Heesterbeek, P.C. de Ruiter, Stability in real food webs: weak links in long loops. Science **296**, 1120–1123 (2002)
24. R. Bagchi, R.E. Gallery, S. Gripenberg et al., Pathogens and insect herbivores drive rainforest plant diversity and composition. Nature **506**, 85–88 (2014)
25. J.R. Britton, Introduced parasites in food webs: new species, shifting structures? Trends Ecol. Evol. **28**, 93–99 (2013)
26. A.P. Dobson, K.D. Lafferty, A.M. Kuris et al., Homage to Linnaeus: how many parasites? How many hosts? PNAS **105**, 11482–11489 (2008)
27. J.A. Dunne, K.D. Lafferty, A.P. Dobson et al., Parasites affect food web structure primarily through increased diversity and complexity. PLoS Biol. **11**, e1001579 (2013)
28. P.J. Hudson, A.P. Dobson, K.D. Lafferty, Is a healthy ecosystem one that is rich in parasites? Trends Ecol. Evol. **21**, 381–385 (2006)
29. K.D. Lafferty, A.P. Dobson, A.M. Kuris, Parasites dominate food web links. PNAS **103**, 11211–11216 (2006)
30. K.D. Lafferty, S. Allesina, M. Arim et al., Parasites in food webs: the ultimate missing links. Ecol. Lett. **11**, 533–546 (2008)
31. S. Selakovic, P.C. de Ruiter, J.A.P. Heesterbeek, Infectious disease agents mediate interaction in food webs and ecosystems. Proc. R. Soc. Sers. B **281**, 20132709 (2014)
32. F. Keesing, L.K. Belden, P. Daszak et al., Impacts of biodiversity on the emergence and transmission of infectious diseases. Nature **468**, 647–652 (2010)
33. M.G. Roberts, J.A.P. Heesterbeek, Characterizing the next-generation matrix and basic reproduction number in ecological epidemiology. J. Math. Biol. **66**, 1045–1064 (2013)
34. O. Diekmann, J.A.P. Heesterbeek, M.G. Roberts, The construction of next-generation matrices for compartmental epidemic systems. J. R. Soc. Interface **7**, 873–885 (2010)
35. O. Diekmann, J.A.P. Heesterbeek, T. Britton, *Mathematical Tools for Understanding Infectious Disease Dynamics* (Princeton University Press, Princeton, 2013)
36. S.B. Carroll, *The Serengeti Rules* (Princeton University Press, Princeton, 2016)
37. M. Buhnerkempe, M.G. Roberts, A.P. Dobson, J.A.P. Heesterbeek et al., Eight challenges in modelling disease ecology in multi-host, multi-agent systems. Epidemics **10**, 26–30 (2015)

Reaction–Diffusion Equations and Cellular Automata

Mikio Murata

Abstract A systematic approach to the construction of cellular automata that are analogs for the reaction–diffusion equations is presented. By using this method for the Allen–Cahn equation and the Gray–Scott model, cellular automatons are constructed. The solutions of the cellular automatons obtained by this method are similar to the solutions of the original reaction–diffusion equations.

Keywords Reaction–diffusion equation · Cellular automaton · Discretization Ultradiscretization

1 Introduction

The reaction–diffusion systems are systems involving constituents locally transformed into each other by chemical reactions and transported in space by diffusion. They arise, quite naturally, in chemistry and chemical engineering but also serve as a reference for the study of a wide range of phenomena encountered beyond the strict realm of chemical science such as environmental and life sciences. For example, the Allen–Cahn equation [1, 2] is a reaction–diffusion equation of mathematical physics which describes the process of phase separation in multicomponent alloy systems, including order–disorder transitions. Reaction and diffusion of chemical species can produce a variety of patterns, reminiscent of those often seen in nature. The Gray–Scott equations [3] model such a reaction. In the Gray–Scott model, various spatiotemporal patterns can be obtained with appropriate parameters. The pattern includes stable standing pulse solutions, traveling pulse solutions, self-replicating patterns, spatiotemporal chaos, and front-wave solutions. A cellular automaton (CA) is a discrete mathematical model studied in computability theory, mathematics, physics, complexity science, theoretical biology, and microstructure

M. Murata (✉)
Institute of Engineering, Tokyo University of Agriculture and Technology, 2-24-16 Nakacho, Koganei-shi, Tokyo 184-8588, Japan
e-mail: mmurata@cc.tuat.ac.jp

R. S. Anderssen et al. (eds.), *Agriculture as a Metaphor for Creativity in All Human Endeavors*, Mathematics for Industry 28,
https://doi.org/10.1007/978-981-10-7811-8_4

modeling. A CA consists of a regular grid of cells, each in one of a finite number of states. For each cell, a set of cells called its neighborhood is defined relative to the specified cell. An initial state is selected by assigning a state for each cell. A new generation is created, according to some fixed rule that determines the new state of each cell in terms of the current state of the cell and the states of the cells in its neighborhood. The rule for updating the state of cells is the same for each cell and does not change over time, and is applied to the whole grid simultaneously. So the reaction–diffusion equation and the cellular automaton are mathematical models having the same role, although their shapes are different. Here, we want to construct a cellular automaton that models the same phenomenon for the phenomena modeled by reaction–diffusion equation. It is convenient if it is possible to construct a cellular automaton from only the form of the reaction–diffusion equation.

Ultradiscretization is one of the methods of formally converting difference equations to piecewise linear equations [4]. Many cellular automata were constructed by applying this method to discrete analogs of soliton equations. By applying this method to solutions of this discrete analog, we have succeeded in constructing solutions of cellular automata as well [5–12]. In addition to the soliton equations, the ultradiscretization can be applied to the discrete analogs of the Painlevé equations. Although it is difficult to clearly describe the solution of the Painlevé equations and its discrete analogs, it is possible to clearly describe the solution of piecewise linear equations obtained by the ultradiscretization [13].

However, in other systems such as reaction–diffusion systems, the ultradiscretization could not be applied. Although cellular automata representing reaction–diffusion phenomena have been proposed in [14], they have not been obtained by the ultradiscretization from the difference equation. The method of constructing the differential equation directly from the cellular automaton is proposed by in [15, 16], but it is not an established method.

We studied a method to construct cellular automata based on the reaction–diffusion equations and succeeded in making difference equations that can be ultradiscretized from the reaction–diffusion equations [17–19]. We applied this method to the Allen–Cahn equation and the Gray–Scott model to construct cellular automata. In addition, it showed that the obtained cellular automata have solutions that behave similarly to the solutions of the original reaction–diffusion equations. In this report, we would like to summarize these results together with new results.

2 Allen–Cahn Cellular Automaton

In this section, a cellular automaton is proposed by employing the tropical discretization [19] for the Allen–Cahn equation. Traveling wave solutions of the one-dimensional cellular automaton are reported.

We now turn to what we consider a systematic approach to discretize differential equations. We call this procedure "the tropical discretization." Let us consider a d-dimensional reaction–diffusion equation

$$\frac{\partial u}{\partial t} = D\Delta u + f(u) - g(u), \tag{1}$$

where $u := u(t, \mathbf{x}), t \geq 0, \mathbf{x} \in \mathbb{R}^d$ and D is positive constant. Δ is the d-dimensional Laplacian $\Delta = \sum_{k=1}^{d} \partial^2/\partial x_k{}^2$. We give a discrete analog of (1). By introducing a function

$$[u_n^{\mathbf{j}}] = \frac{1}{2d} \sum_{k=1}^{d} \left(u_n^{\mathbf{j}+\mathbf{e}_k} + u_n^{\mathbf{j}-\mathbf{e}_k} \right), \tag{2}$$

where $u(n, \mathbf{j})(=: u_n^{\mathbf{j}}) : \mathbb{N} \times \mathbb{Z}^d \to \mathbb{R}$ and where $\mathbf{e}_k \in \mathbb{Z}^d$ is a unit vector whose kth component is 1 and whose other components are 0. We construct a partial difference equation

$$u_{n+1}^{\mathbf{j}} = [u_n^{\mathbf{j}}] \frac{\varepsilon^{-1}[u_n^{\mathbf{j}}] + f([u_n^{\mathbf{j}}])}{\varepsilon^{-1}[u_n^{\mathbf{j}}] + g([u_n^{\mathbf{j}}])}, \tag{3}$$

where ε is a positive constant. This equation is a discretization of (1), because (3) is equivalent to

$$\frac{u_{n+1}^{\mathbf{j}} - u_n^{\mathbf{j}}}{\varepsilon} = \frac{\delta^2}{2d\varepsilon} \sum_{k=1}^{d} \frac{u_n^{\mathbf{j}+\mathbf{e}_k} + u_n^{\mathbf{j}-\mathbf{e}_k} - 2u_n^{\mathbf{j}}}{\delta^2} + \frac{[u_n^{\mathbf{j}}] \left\{ f([u_n^{\mathbf{j}}]) - g([u_n^{\mathbf{j}}]) \right\}}{[u_n^{\mathbf{j}}] + \varepsilon g([u_n^{\mathbf{j}}])}.$$

if there exists smooth functions $u(t, \mathbf{x})$ $(t \geq 0,\ \mathbf{x} \in \mathbb{R}^d)$ that satisfy $u(\delta n, \varepsilon \mathbf{j}) = u_n^{\mathbf{j}}$, the d-dimensional reaction–diffusion equation (1) are obtained with the limit $\varepsilon \to 0$.

The Allen–Cahn equation [1] is the partial differential equation

$$\frac{\partial u}{\partial t} = D\Delta u - (u-1)(u-b)(u-a), \quad 1 < b < a, \tag{4}$$

where zero points of reaction term are $u = 1,\ b,\ a$. Equation (4) appears in [2]. Equation (4) can be called the unbalanced Allen–Cahn equation, and it is also called the Nagumo equation. We apply the tropical discretization to (4). Considering $f(u) = (1 + a + b)\, u^2 + ab$ and $g(u) = u^3 + (a + b + ab)\, u$ in (1) and putting $f(u)$ and $g(u)$ to (3), then we have a difference equation

$$u_{n+1}^{\mathbf{j}} = \frac{\varepsilon^{-1}[u_n^{\mathbf{j}}] + (1 + a + b)\, [u_n^{\mathbf{j}}]^2 + ab}{\varepsilon^{-1} + [u_n^{\mathbf{j}}]^2 + a + b + ab}. \tag{5}$$

Next, we construct a piecewise linear equation from (5) by using the ultradiscretization procedure [4]. The procedure of ultradiscretization is equivalent to the construction of a piecewise linear equation using the following substitution rule.

Let

$$\varepsilon \to E,\ a \to A,\ b \to B,\ u_n^{\mathbf{j}} \to U_n^{\mathbf{j}}$$

Discrete equation	Ultradiscrete equation
$\times$ (Multiplication)	$+$ (Addition)
$\div$ (Division)	$-$ (Subtraction)
$+$ (Addition)	max (Maximum)
$a, b, c, \ldots$ (Positive value variables)	$A, B, C, \ldots$ (Real value variables)
$1, \frac{1}{2}, \sqrt{2}, \ldots$ (Positive constants)	0 (Zero constant)

and applying this transformation rule to (2) and (5) yields the following function:

$$\{U_n^{\mathbf{j}}\} = \max_{k=1,\ldots,d} \left(U_n^{\mathbf{j}+\mathbf{e}_k}, U_n^{\mathbf{j}-\mathbf{e}_k}\right),$$

and an ultradiscrete analog of Allen–Cahn equation

$$\begin{aligned} U_{n+1}^{\mathbf{j}} = \max &\left\{\{U_n^{\mathbf{j}}\} - E, \max(0, A, B) + 2\{U_n^{\mathbf{j}}\}, A + B\right\} \\ &- \max\left\{-E, 2\{U_n^{\mathbf{j}}\}, A, B, A + B\right\}. \end{aligned} \tag{6}$$

We deal with the case of $0 < B < A$ since $1 < b < a$. From now on, we deal with only the case of $E > -A - B$. Then, (6) becomes

$$U_{n+1}^{\mathbf{j}} = \max\left\{\{U_n^{\mathbf{j}}\} - A - E, 2\{U_n^{\mathbf{j}}\}, B\right\} - \max\left\{2\{U_n^{\mathbf{j}}\} - A, B\right\}. \tag{7}$$

When $-A - B < E < -A - \frac{B}{2}$, (7) is equivalent to

$$U_{n+1}^{\mathbf{j}} = \begin{cases} 0 & (\{U_n^{\mathbf{j}}\} < A + B + E) \\ \{U_n^{\mathbf{j}}\} - A - B - E & (A + B + E \le \{U_n^{\mathbf{j}}\} < \frac{B}{2}) \\ 2\{U_n^{\mathbf{j}}\} - B & (\frac{B}{2} \le \{U_n^{\mathbf{j}}\} < \frac{A+B}{2}) \\ A & (\frac{A+B}{2} \le \{U_n^{\mathbf{j}}\}). \end{cases}$$

When $E \ge -A - \frac{B}{2}$, (7) is equivalent to

$$U_{n+1}^{\mathbf{j}} = \begin{cases} 0 & (\{U_n^{\mathbf{j}}\} < \frac{B}{2}) \\ 2\{U_n^{\mathbf{j}}\} - B & (\frac{B}{2} \le \{U_n^{\mathbf{j}}\} < \frac{A+B}{2}) \\ A & (\frac{A+B}{2} \le \{U_n^{\mathbf{j}}\}). \end{cases}$$

Let $A, B, E \in \mathbb{Z}$ and initial data of (7) $U_n^{\mathbf{j}} \in \{0, 1, \ldots, A\}$. The solution of (7) becomes a cellular automaton with $A + 1$-states. For example, the rule table for $A = 4$, $B = 2$ is

$\{U_n^{\mathbf{j}}\}$	4	3	2	1	0
$U_{n+1}^{\mathbf{j}}$	4	4	2	0	0

Now, let spatial dimension $d = 1$. For the one-dimensional cellular automaton (7), several traveling solutions exist. By using the step function

$$H(x) = \begin{cases} 1 & (x \geq 0) \\ 0 & (x < 0), \end{cases} \tag{8}$$

$$U_n^j = A\,H(n - j) \tag{9}$$

is a front traveling solution for wave speed 1 that satisfies $\lim_{j\to-\infty} U_n^j = A$ and $\lim_{j\to\infty} U_n^j = 0$.

$$U_n^j = B\,H(n - j) \tag{10}$$

is a front traveling solution for wave speed 1 that satisfies $\lim_{j\to-\infty} U_n^j = B$ and $\lim_{j\to\infty} U_n^j = 0$. The superposition of (9) and (10)

$$U_n^j = \max\{A\,H(n - j),\, B\,H(n + n_0 - j)\} \tag{11}$$

is a front traveling solution for wave speed 1 that satisfies $\lim_{j\to-\infty} U_n^j = A$ and $\lim_{j\to\infty} U_n^j = 0$.

$$U_n^j = \max\{A\,H(n - j),\, B\} \tag{12}$$

is also a front traveling solution for wave speed 1 that satisfies $\lim_{j\to-\infty} U_n^j = A$ and $\lim_{j\to\infty} U_n^j = B$. If U_n^j is a solution of (7), then U_n^{-j} is a solution of (7). Therefore, front traveling solutions for every wave speed -1 also exist. For the real-valued equation of (7), many traveling solutions for every wave speed exist [19]. If a wave traveling to the right collides with a wave traveling to the left, then the two traveling waves disappear, see also Fig. 1.

Let spatial dimension $d = 2$. Let us consider the 2D Allen–Cahn CA with $A = 4$, $B = 2$ as a mathematical model of forest fires. We can assign a numerical value to each cell following this convention:

- The state where it can be ignited has a value of 0,
- A smoldering cell has a value of 1,

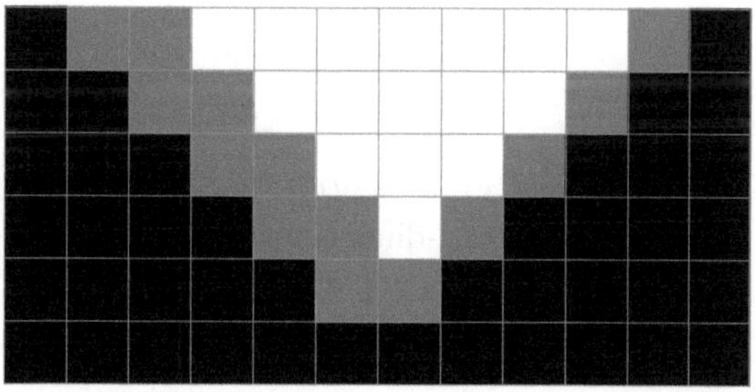

Fig. 1 A collision and a disappearance of two traveling waves in Allen–Cahn cellular automaton with $A = 4$, $B = 2$. Values of U_n^j are represented as follows: 0 (white), 2 (gray), and 4 (black). Time progresses vertically from top to bottom

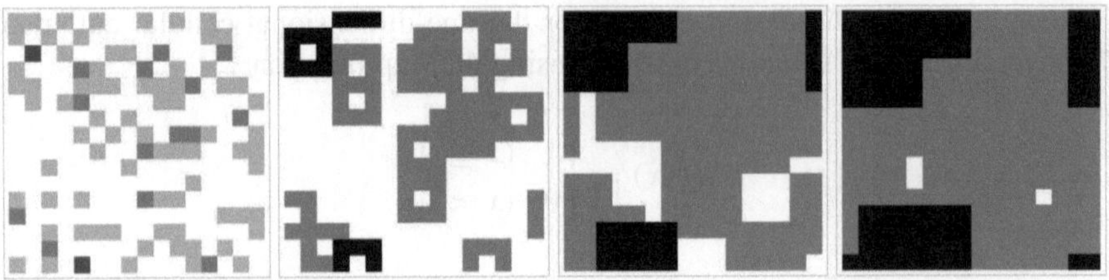

Fig. 2 2D Allen–Cahn CA with $A = 4$, $B = 2$. Values of U_n^j are represented as follows: 0 (white), 2 (gray), and 4 (black)

- A burning cell has a value of 2,
- A being extinguished cell has a value of 3, and
- A burnt cell has a value of 4.

The pattern when starting from an appropriate initial value is shown in Fig. 2. We can see how the fire burns and then the fire is extinguished. Of course, this cellular automaton is a simple model. In order to apply it to real problems, we think that it is better to improve based on this model.

3 Gray–Scott Cellular Automaton

Gray–Scott model [3] is a variant of the autocatalytic model. Basically, it considers the reactions

$$\mathrm{U} + 2\mathrm{V} \to 3\mathrm{V},$$
$$\mathrm{V} \to \mathrm{P},$$

in an open flow reactor where U is continuously supplied, and the product P removed.

A mathematical model of the reactions above is the following system of partial differential equations:

$$\frac{\partial u}{\partial t} = D_u \Delta u - uv^2 + a(1-u), \tag{13a}$$

$$\frac{\partial v}{\partial t} = D_v \Delta v + uv^2 - bv, \tag{13b}$$

where $u := u(t, \mathbf{x})$, $v := v(t, \mathbf{x})$, $t \geq 0$, $\mathbf{x} \in \mathbb{R}^d$, and D_u, D_v, a, and b are positive constants. Δ is the d-dimensional Laplacian. The solutions of this system represent spatial patterns. Changing not only an initial condition but also parameters, various patterns are observed [20–22].

In this section, we discretize and ultradiscretize (13) and investigate solutions. Since it is more convenient to consider the ultradiscretization, we take the scaling $w := v + 1$ which changes (13) to

$$\frac{\partial u}{\partial t} = D_u \Delta u - u(w-1)^2 + a(1-u), \tag{14a}$$

$$\frac{\partial w}{\partial t} = D_v \Delta w + u(w-1)^2 - b(w-1). \tag{14b}$$

Let us consider two-component d-dimensional reaction–diffusion equations

$$\frac{\partial u}{\partial t} = D_u \Delta u + f_u(u,w) - g_u(u,w), \tag{15a}$$

$$\frac{\partial w}{\partial t} = D_w \Delta w + f_w(u,w) - g_w(u,w). \tag{15b}$$

We construct partial difference equations

$$u_{n+1}^{\mathbf{j}} = [u_n^{\mathbf{j}}]_p \frac{\varepsilon^{-1}[u_n^{\mathbf{j}}]_p + f_u([u_n^{\mathbf{j}}]_p, w_{n+1}^{\mathbf{j}})}{\varepsilon^{-1}[u_n^{\mathbf{j}}]_p + g_u([u_n^{\mathbf{j}}]_p, w_{n+1}^{\mathbf{j}})}, \tag{16a}$$

$$w_{n+1}^{\mathbf{j}} = [w_n^{\mathbf{j}}]_q \frac{\varepsilon^{-1}[w_n^{\mathbf{j}}]_q + f_w([u_n^{\mathbf{j}}]_p, [w_n^{\mathbf{j}}]_q)}{\varepsilon^{-1}[w_n^{\mathbf{j}}]_q + g_w([u_n^{\mathbf{j}}]_p, [w_n^{\mathbf{j}}]_q)}, \tag{16b}$$

where $n \in \mathbb{Z}_{\geq 0}$, $\mathbf{j} \in \mathbb{Z}^d$ and

$$[u_n^{\mathbf{j}}]_p := \sum_{k=1}^{d} \frac{u_n^{\mathbf{j}+p\mathbf{e}_k} + u_n^{\mathbf{j}-p\mathbf{e}_k}}{2d} \quad (p \in \mathbb{N}),$$

$$[w_n^{\mathbf{j}}]_q := \sum_{k=1}^{d} \frac{w_n^{\mathbf{j}+q\mathbf{e}_k} + w_n^{\mathbf{j}-q\mathbf{e}_k}}{2d} \quad (q \in \mathbb{N}).$$

This equation is a discretization of (15). Considering $f_u(u,w) = 2uw + a$, $g_u(u,w) = uw^2 + (a+1)u$, $f_w(u,w) = u(w^2+1) + b$ and $g_w(u,w) = (2u+b)w$ in (15) and putting them to (16), then we have a system of partial difference equations

$$u_{n+1}^{\mathbf{j}} = \frac{\varepsilon^{-1}[u_n^{\mathbf{j}}]_p + 2[u_n^{\mathbf{j}}]_p w_{n+1}^{\mathbf{j}} + a}{\varepsilon^{-1} + (w_{n+1}^{\mathbf{j}})^2 + a + 1}, \tag{17a}$$

$$w_{n+1}^{\mathbf{j}} = \frac{\varepsilon^{-1}[w_n^{\mathbf{j}}]_q + [u_n^{\mathbf{j}}]_p\{([w_n^{\mathbf{j}}]_q)^2 + 1\} + b}{\varepsilon^{-1} + 2[u_n^{\mathbf{j}}]_p + b}. \tag{17b}$$

We ultradiscretize (17) and investigate the solutions. Let

$$\varepsilon \to E,\ a \to A,\ b \to B,\ u_n^{\mathbf{j}} \to U_n^{\mathbf{j}},\ w_n^{\mathbf{j}} \to W_n^{\mathbf{j}}$$

and applying the substitution rule for the ultradiscretization, then we have

$$U_{n+1}^{\mathbf{j}} = \max(\{U_n^{\mathbf{j}}\}_p - E, \{U_n^{\mathbf{j}}\}_p + W_{n+1}^{\mathbf{j}}, A) - \max(-E, 2W_{n+1}^{\mathbf{j}}, A, 0), \tag{18a}$$

$$W_{n+1}^{\mathbf{j}} = \max(\{W_n^{\mathbf{j}}\}_q - E, \{U_n^{\mathbf{j}}\}_p + 2\{W_n^{\mathbf{j}}\}_q, \{U_n^{\mathbf{j}}\}_p, B) - \max(-E, \{U_n^{\mathbf{j}}\}_p, B), \tag{18b}$$

where

$$\{U_n^{\mathbf{j}}\}_p := \max_{k=1,\dots,d}(U_n^{\mathbf{j}+p\mathbf{e}_k}, U_n^{\mathbf{j}-p\mathbf{e}_k}),$$

$$\{W_n^{\mathbf{j}}\}_q := \max_{k=1,\dots,d}(W_n^{\mathbf{j}+q\mathbf{e}_k}, W_n^{\mathbf{j}-q\mathbf{e}_k}).$$

Taking a limit $E \to \infty$ and assuming $W_n^{\mathbf{j}} \geq 0$ in (18), then we get

$$U_{n+1}^{\mathbf{j}} = \max(\{U_n^{\mathbf{j}}\}_p + W_{n+1}^{\mathbf{j}}, A) - \max(2W_{n+1}^{\mathbf{j}}, A), \tag{19a}$$

$$W_{n+1}^{\mathbf{j}} = \max(\{U_n^{\mathbf{j}}\}_p + 2\{W_n^{\mathbf{j}}\}_q, B) - \max(\{U_n^{\mathbf{j}}\}_p, B). \tag{19b}$$

Let spatial dimension $d = 1$ and initial data of (19) $U_0^j \in \{0, -1\}$, $W_0^j \in \{0, 1\}$. Taking some conditions to parameters A and B, the solution of (19) becomes to a cellular automaton. There are five types of conditions for A and B as follows:

Type I	Type II	Type III	Type IV	Type V
$A \leq -1$	$0 \leq A \leq 1$	$A \geq 2$	$A \leq -1$	$A \geq 0$
$B = 1$	$B = 1$	$B = 1$	$B \geq 2$	$B \geq 2$

If we take the scaling $X_n^j = U_n^j + 1$, then $X_n^j \in \{0, 1\}$ and $W_n^j \in \{0, 1\}$.
Type I: The rule for $A \leq -1$, $B = 1$:

$\{X_n^j\}_p, \{W_n^j\}_q$	1, 1	1, 0	0, 1	0, 0
X_{n+1}^j	0	1	0	0
W_{n+1}^j	1	0	0	0

In this case, we can observe the patterns in Fig. 3. Values of W_n^j are represented as follows: 0 (white) and 1 (black). Different patterns arise due to differences in diffusion parameters p and q. A surviving standing wave is observed in the case of $(p, q) = (1, 0)$. When time passes for a while, the state becomes uniform in the case of $(p, q) = (0, 1)$. Collision and extinction of traveling waves are observed in the case of $(p, q) = (1, 1)$ and $(p, q) = (2, 1)$.
Type II: The rule for $0 \leq A \leq 1$, $B = 1$:

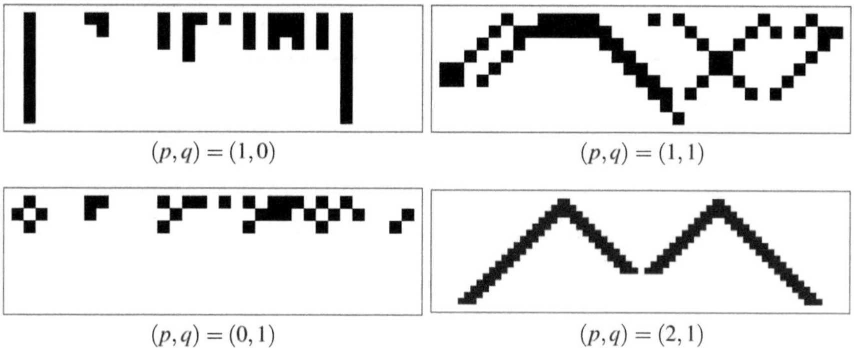

Fig. 3 Spatiotemporal patterns of Type I

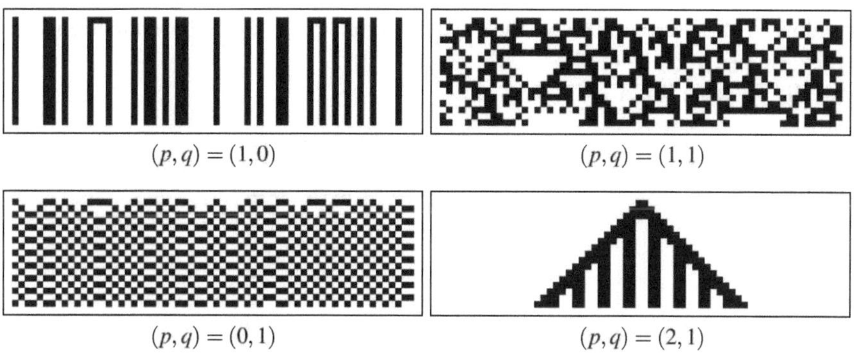

Fig. 4 Spatiotemporal patterns of Type II

$\{X_n^j\}_p, \{W_n^j\}_q$	1, 1	1, 0	0, 1	0, 0
X_{n+1}^j	0	1	1	1
W_{n+1}^j	1	0	0	0

We can observe the spatiotemporal patterns of Type II in Fig. 4. Since this relation $X_{n+1}^j = 1 - W_{n+1}^j$ is held, W_n^j satisfies a single equation. Moreover, taking $(p, q) = (1, 1)$, the equation is same as Elementary cellular automaton (ECA) rule 90, which is well known for fractal design:

$W_n^{j-1}\ W_n^j\ W_n^{j+1}$	111	110	101	100	011	010	001	000
W_{n+1}^j	0	1	0	1	1	0	1	0

taking $(p, q) = (1, 0)$, the equation is same as ECA rule 76:

$W_n^{j-1}\ W_n^j\ W_n^{j+1}$	111	110	101	100	011	010	001	000
W_{n+1}^j	0	1	0	0	1	1	0	0

and taking $(p, q) = (0, 1)$, the equation is same as ECA rule 50:

$W_n^{j-1}\ W_n^j\ W_n^{j+1}$	111	110	101	100	011	010	001	000
W_{n+1}^j	0	0	1	1	0	0	1	0

Pattern changes dramatically with changes in diffusion parameters p and q. There are several standing waves in the case of $(p, q) = (1, 0)$. Vibration patterns are observed in the case of $(p, q) = (0, 1)$. Its time–space diagram forms many triangular "windows" of different sizes in the case of $(p, q) = (1, 1)$. Self-replication occurs in the case of $(p, q) = (2, 1)$.
Type III: The rule for $A \geq 2$, $B = 1$:

$\{X_n^j\}_p, \{W_n^j\}_q$	1, 1	1, 0	0, 1	0, 0
X_{n+1}^j	1	1	1	1
W_{n+1}^j	1	0	0	0

In this case, $X_{n+1}^j = 1$ so that W_n^j satisfies $W_{n+1}^j = \{W_n^j\}_q$.
Type IV: The rule of $A \leq -1$, $B \geq 2$:

$\{X_n^j\}_p, \{W_n^j\}_q$	1, 1	1, 0	0, 1	0, 0
X_{n+1}^j	1	1	0	0
W_{n+1}^j	0	0	0	0

In this case, $W_{n+1}^j = 0$ so that X_n^j satisfies $X_{n+1}^j = \{X_n^j\}_p$.
Type V: The rule of $A \geq 0$, $B \geq 2$:

$\{X_n^j\}_p, \{W_n^j\}_q$	1, 1	1, 0	0, 1	0, 0
X_{n+1}^j	1	1	1	1
W_{n+1}^j	0	0	0	0

In this case, $X_{n+1}^j = 1$ and $W_{n+1}^j = 0$ so that X and W do not change in future.

If one take $B \geq L$, $U_n^j \in \{0, -1, \ldots, -L\}$ and $W_n^j \in \{0, 1, \ldots, L\}$, the solution of (19) becomes to a cellular automaton whose dependent variable can have $L + 1$

values. The more L is large, the more the number of rule for evolution. In this case, the spatial pattern is also classified five types as follows:

Type I	Type II	Type III	Type IV	Type V
$A \le -1$	$0 \le A \le 2L-1$	$A \ge 2L$	$A \le -1$	$A \ge 0$
$B = L$	$B = L$	$B = L$	$B \ge L+1$	$B \ge L+1$

Now, let spatial dimension $d = 2$. We also take similar condition to the initial condition of (19) in the case of $d = 1$: $U_0^{\mathbf{j}} \in \{0, -1\}$, $W_0^{\mathbf{j}} \in \{0, 1\}$. We can separate spatial patterns to five types as similar to the case of $d = 1$. If $A \le -1$, $B = 1$, then the rule of the evolution is as follows:

$\{X_n^{\mathbf{j}}\}_p, \{W_n^{\mathbf{j}}\}_q$	1, 1	1, 0	0, 1	0, 0
$X_{n+1}^{\mathbf{j}}$	0	1	0	0
$W_{n+1}^{\mathbf{j}}$	1	0	0	0

In this case, the pattern in Fig. 5 is observed. Values of $W_n^{\mathbf{j}}$ are represented as follows: 0 (white) and 1 (black). In Fig. 5, the solution of (19) with $A = -1$, $B = 1$ gives a ring pattern. The ring grows over time.

If $0 \le A \le 1$, $B = 1$, the rule of evolution is as follows:

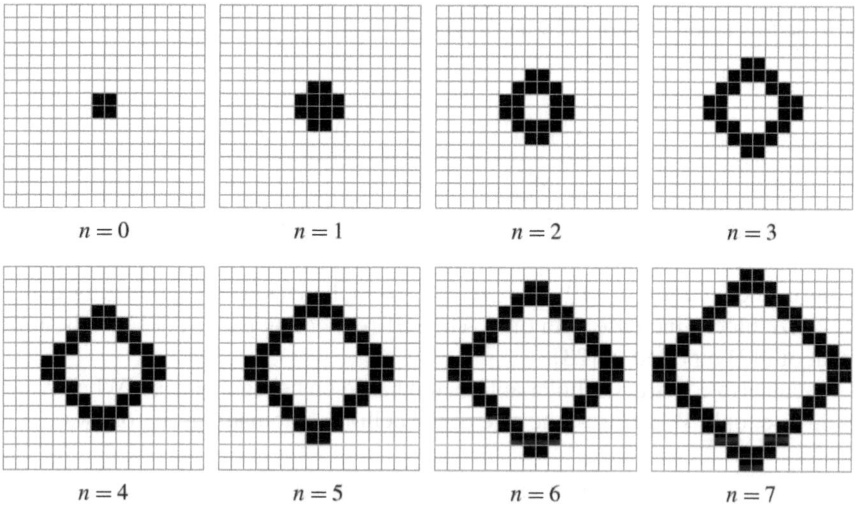

Fig. 5 Ring pattern. $W_n^{\mathbf{j}}$ with $(p, q) = (1, 1)$

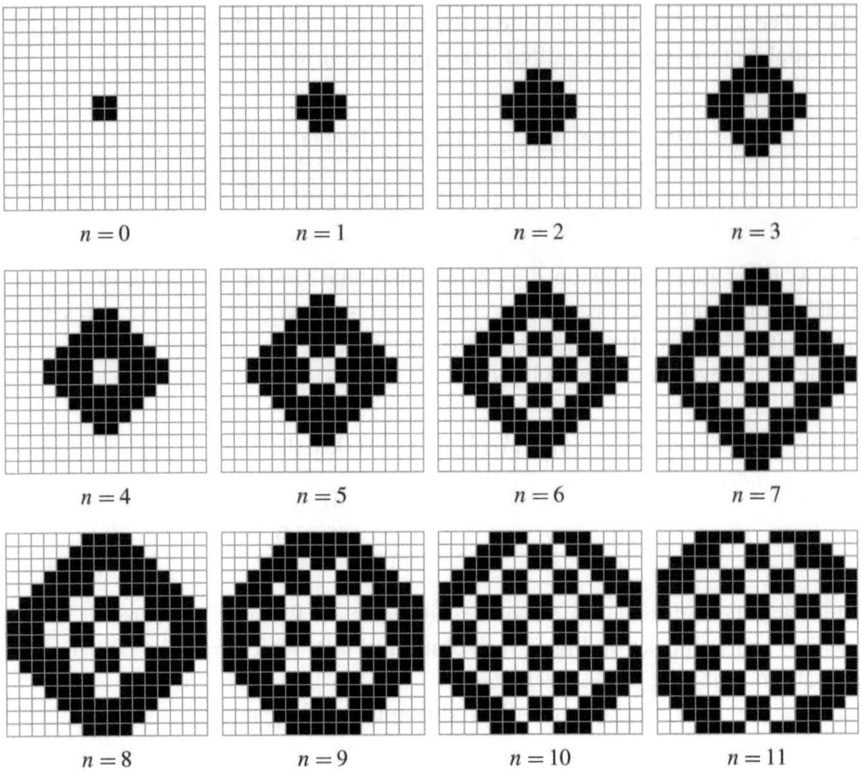

Fig. 6 Self-replicating pattern. $W_n^{\mathbf{j}}$ with $(p, q) = (2, 1)$

$\{X_n^{\mathbf{j}}\}_p, \{W_n^{\mathbf{j}}\}_q$	1, 1	1, 0	0, 1	0, 0
$X_{n+1}^{\mathbf{j}}$	0	1	1	1
$W_{n+1}^{\mathbf{j}}$	1	0	0	0

In Fig. 6, the solution of (19) with $A = 0,\ B = 1$ gives a self-replicating pattern. 2×2-size blocks increase with time. The self-replicating pattern of Fig. 6 is similar to that of (13) reported in [23].

We can also construct cellular automaton based on hexagonal cells. When X does not diffuse, only W diffuses, the rule of the cellular automaton is as follows:

$X_n^{j,k}, \{W_n^{j,k}\}$	1, 1	1, 0	0, 1	0, 0
$X_{n+1}^{j,k}$	0	1	1	1
$W_{n+1}^{j,k}$	1	0	0	0

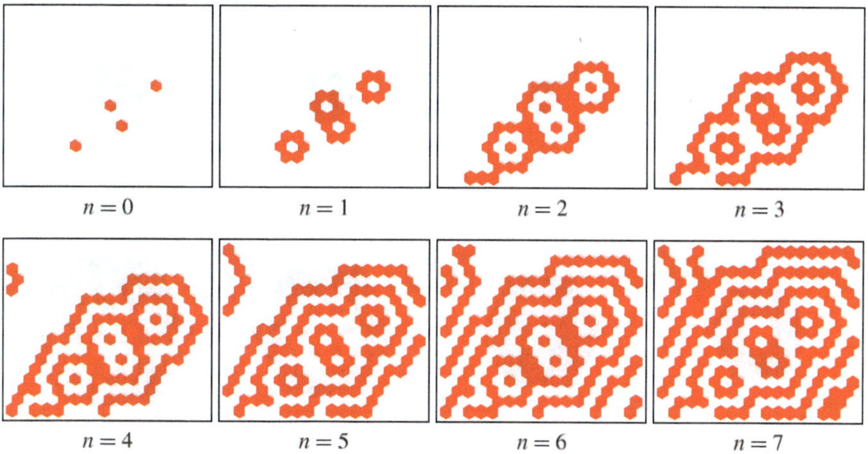

Fig. 7 Target pattern of the cellular automaton based on hexagonal cells

where $\{W_n^{j,k}\} = \max(W_n^{j+1,k}, W_n^{j-1,k}, W_n^{j,k+1}, W_n^{j,k-1}, W_n^{j+1,k+1}, W_n^{j-1,k-1})$. Target pattern that resembles those in the Belousov–Zhabotinsky reaction which is a spatiotemporal chemical oscillator is observed in Fig. 7.

As shown in Fig. 8, a self-replicating pattern appears even in the following cellular automaton on hexagonal cells:

$\{X_n^{j,k}\}_2, \{W_n^{j,k}\}_1$	1, 1	1, 0	0, 1	0, 0
$X_{n+1}^{j,k}$	0	1	1	1
$W_{n+1}^{j,k}$	1	0	0	0

where

$$\{X_n^{j,k}\}_2 = \max(X_n^{j+2,k}, X_n^{j-2,k}, X_n^{j,k+2}, X_n^{j,k-2}, X_n^{j+2,k+2}, X_n^{j-2,k-2}),$$
$$\{W_n^{j,k}\}_1 = \max(W_n^{j+1,k}, W_n^{j-1,k}, W_n^{j,k+1}, W_n^{j,k-1}, W_n^{j+1,k+1}, W_n^{j-1,k-1}).$$

Gray–Scott model is a model based on a chemical reaction, but a reaction–diffusion system can construct a mathematical model of various phenomena by a difference in reaction term. So, this cellular automaton is also considered to be a mathematical model of various phenomena by changing the function expressing the reaction.

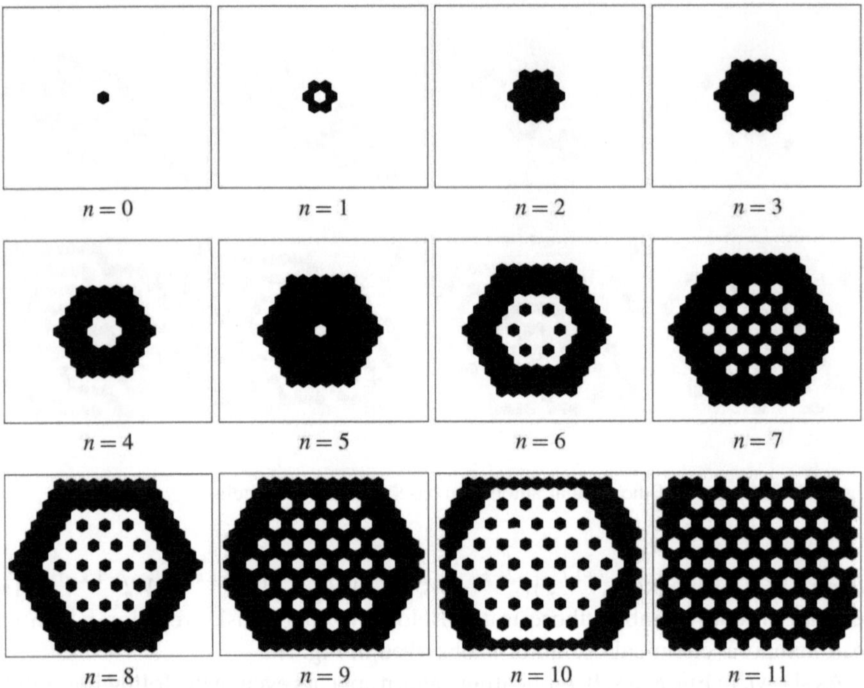

Fig. 8 Self-replicating pattern of the cellular automaton based on hexagonal cells

4 Concluding Remarks

In this article, we proposed and investigated the cellular automata that are analogs for Allen–Cahn equation and Gray–Scott model, which are the reaction–diffusion systems. The solutions of the cellular automata have patterns similar to the solution of the differential equations. The author expects that characteristics of solutions for cellular automata will provide guidance for clarifying characteristics of solutions for differential equations. In 1985, Wolfram listed up 20 important problems in the research of CAs [24]. The ninth problem asks "What is the correspondence between cellular automata and continuous systems?" He pointed out the similarity between time evolution patterns of CAs and behavior of continuous systems described by differential equations and stated that discretization of time and spatial valuables would correspond to an approximation in numerical calculation, but that the meaning of a discretization of physical quantities is not clear. Concerning this problem, a systematic method to construct CAs from reaction–diffusion systems (nonintegrable systems) has been proposed in this paper.

Acknowledgements This work was supported by JSPS KAKENHI Grant Number 16K21024.

References

1. S. Allen, J.W. Cahn, A microscopic theory for antiphase boundary motion and its application to antiphase domain coarsening. Acta. Metall. **27**, 1084–1095 (1979)
2. J. Nagumo, S. Yoshizawa, S. Arimoto, Bistable transmission lines. IEEE Trans. Circuit Theory **CT-12**(3) (1965)
3. P. Gray, S.K. Scott, Sustained oscillations and other exotic patterns of behaviour in isothermal reactions. J. Phys. Chem. **89**, 22–32 (1985)
4. T. Tokihiro, D. Takahashi, J. Matsukidaira, J. Satsuma, From soliton equations to integrable cellular automata through a limiting procedure. Phys. Rev. Lett. **29**, 3247–3250 (1996)
5. M.J. Ablowitz, J.M. Keiser, L.A. Takhtajan, Stable, multi-state, time-reversible cellular automata with rich particle content. Quaest. Math. **15**, 325–343 (1992)
6. A.S. Fokas, E. Papadopoulou, Y. Saridakis, Soliton cellular automata. Phys. D **41**, 297–321 (1990)
7. A.S. Fokas, E. Papadopoulou, Y. Saridakis, M.J. Ablowitz, Interaction of simple particles in soliton cellular automata. Stud. Appl. Math. **81**, 153–180 (1989)
8. J. Matsukidaira, J. Satsuma, D. Takahashi, T. Tokihiro, M. Torii, Toda-type cellular automaton and its N-soliton solution. Phys. Lett. A **225**, 287–295 (1997)
9. M. Murata, S. Isojima, A. Nobe, J. Satsuma, Exact solutions for discrete and ultradiscrete modified KdV equations and their relation to box-ball systems. J. Phys. A Math. Gen. **39**, L27–L34 (2006)
10. A. Nagai, D. Takahashi, T. Tokihiro, Soliton cellular automaton, Toda molecule equation and sorting algorithm. Phys. Lett. A **255**, 265–271 (1999)
11. J.K. Park, K. Steiglitz, W.P. Thurston, Soliton-like behavior in automata. Phys. D **19**, 423–432 (1986)
12. D. Takahashi, J. Satsuma, A soliton cellular automaton. J. Phys. Soc. Jpn. **59**, 3514–3519 (1990)
13. M. Murata, Exact solutions with two parameters for an ultradiscrete Painlevé equation of type $A_6^{(1)}$. SIGMA, **7** (2011), 059, 15pp
14. D. Takahashi, A. Shida, M. Usami, On the pattern formation mechanism of (2+1)D max-plus models. J. Phys. A Math. Gen. **34**, 10715–10726 (2001)
15. W. Kunishima, A. Nishiyama, H. Tanaka, T. Tokihiro, Differential equations for creating complex cellular automaton patterns. J. Phys. Soc. Jpn. **73**, 2033–2036 (2004)
16. H. Tanaka, A. Nakajima, A. Nishiyama, T. Tokihiro, Derivation of a differential equation exhibiting replicative time-evolution patterns by inverse ultra-discretization. J. Phys. Soc. Jpn. **78** (2009), 034002, 5pp
17. K. Matsuya, M. Murata, Spatial pattern of discrete and ultradiscrete Gray-Scott model. Discrete Contin. Dyn. Syst. Ser. B **20**, 173–187 (2015)
18. M. Murata, Multidimensional traveling waves in Allen–Cahn cellular automaton. J. Phys. A Math. Theor. **48** (2015), 255202, 10pp
19. M. Murata, Tropical discretization: ultradiscrete Fisher-KPP equation and ultradiscrete Allen-Cahn equation. J. Difference. Equ. Appl. **19**, 1008–1021 (2013)
20. W. Mazin, K.E. Rasmussen, E. Mosekilde, P. Borckmans, G. Dewel, Pattern formation in the bistable Gray-Scott model. Math. Comput. Simul. **40**, 371–396 (1996)
21. Y. Nishiura, D. Ueyama, A skeleton structure of self-replicating dynamics. Phys. D **130**, 73–104 (1999)
22. Y. Nishiura, D. Ueyama, Spatio-temporal chaos for the Gray-Scott model. Phys. D **150**, 137–162 (2001)
23. J.E. Pearson, Complex patterns in a simple system. Science **261**, 189–192 (1993)
24. S. Wolfram, Twenty problems in the theory of cellular automata. Phys. Scripta. **1985**, 170–183 (1985)

Geometry of Timelike Minimal Surfaces and Null Curves

Shintaro Akamine

Abstract In this chapter, we investigate the behavior of the Gaussian curvature of timelike minimal surfaces with or without singular points in the 3-dimensional Lorentz–Minkowski space. For timelike minimal surfaces without singular points, we prove that the sign of the Gaussian curvature, which corresponds to diagonalizability of the shape operator, of any timelike minimal surface is determined by the degeneracy and the orientations of the two null curves that generate the surface. Moreover, we also determine the behavior of the Gaussian curvature near cuspidal edges, swallowtails, and cuspidal cross caps on timelike minimal surfaces. We show that there are no umbilic points near cuspidal edges on a timelike minimal surface. Near swallowtails, we show that the sign of the Gaussian curvature is negative, that is, we can take always real principal curvatures near swallowtails. Near cuspidal cross caps, we also show that the sign of the Gaussian curvature is positive, that is, we can take only complex principal curvatures near cuspidal cross caps.

Keywords Lorentz–Minkowski space · Timelike minimal surface · Gaussian curvature · Singularity

1 Introduction

The shape of an equilibrium liquid crystal is obtained by minimizing an anisotropic energy. As pointed out by Chandrasekhar [3], when the anisotropy is large, the equilibrium shape will be like a tactoid (see Fig. 2b in [3]) and it is known that such shapes were observed in plant virus preparations in Biology. On the other hand, Honda, Koiso, and Tanaka [7] recently proved that surfaces with zero mean curvature in the 3-dimensional Lorentz-Minkowski space $\mathbb{R}^3_1$ arise as stationary surfaces for an anisotropic energy, and some timelike surfaces also have shapes of tactoids

S. Akamine (✉)
Graduate School of Mathematics, Kyushu University, 744 Motooka,
Nishi-ku, Fukuoka 819-0395, Japan
e-mail: s-akamine@math.kyushu-u.ac.jp

R. S. Anderssen et al. (eds.), *Agriculture as a Metaphor for Creativity in All Human Endeavors*, Mathematics for Industry 28,
https://doi.org/10.1007/978-981-10-7811-8_5

(see Fig. 3 in [7]). These timelike minimal surfaces are not energy minimizers for the anisotropic energy introduced in [7]. However, since these surfaces are closed and convex, there exist anisotropic energy functionals for which they are energy minimizers among closed surfaces enclosing the same volume (cf. [17]). Hence, the anisotropic energy for which a tactoid is a stationary surface is not unique. Therefore, we can determine neither the anisotropic energy by looking at shapes of tactoids (or plant virus preparations) nor all shapes of stationary surfaces for these e nergies. In this chapter, we study the behavior of the Gaussian curvature of timelike minimal surfaces, which represents geometric shapes of stationary surfaces for an anisotropic energy used in [7].

A timelike (resp. spacelike) surface in $\mathbb{R}^3_1$ is a surface whose first fundamental form is a Lorentzian (resp. Riemannian) metric. One of the most important differences between spacelike surfaces in $\mathbb{R}^3_1$ (or surfaces in 3-dimensional Euclidean space $\mathbb{E}^3$) and timelike surfaces in $\mathbb{R}^3_1$ is the diagonalizability of the shape operator, that is, the shape operator of a timelike surface is not always diagonalizable over real number field $\mathbb{R}$. The characteristic equation of the shape operator of any surface in $\mathbb{E}^3$, spacelike or timelike surface in $\mathbb{R}^3_1$ can be written as $\lambda^2 - 2H\lambda + K = 0$, where H and K are the mean curvature and Gaussian curvature of the surface. Hence, its discriminant can be written $H^2 - K$. In this chapter, we discuss surfaces whose mean curvature H vanishes identically. A surface whose mean curvature vanishes identically is called a *zero mean curvature surface* (ZMC surface for short).

A ZMC surface in $\mathbb{E}^3$ is called a *minimal surface*, which is a critical point of the area functional on the space of all surfaces having the same boundary. A minimal surface in $\mathbb{E}^3$ is a mathematical model of a soap film. A spacelike ZMC surface, which is called a *maximal surface* is not only a critical point but also a local maximizer of the area functional. Since shape operators of these surfaces are diagonalizable over $\mathbb{R}$, their Gaussian curvatures are always non-positive.

On the other hand, a timelike ZMC surface, which is called a *timelike minimal surface* neither maximize nor minimize the area functional. Timelike minimal surfaces have been also studied from the viewpoint of the string theory, see, for example, [2, 8, 11] and their references. As mentioned above, in contrast to minimal surfaces in $\mathbb{E}^3$ and maximal surfaces in $\mathbb{R}^3_1$, the shape operator of a timelike minimal surface is not always diagonalizable. The diagonalizability of the shape operator S of a timelike minimal surface is determined by the sign of the Gaussian curvature. The shape operator S is diagonalizable over $\mathbb{R}$ on points with negative Gaussian curvature, S is diagonalizable over the complex number field $\mathbb{C}$ on points with positive Gaussian curvature, and flat points consist of umbilic points and quasi-umbilic points. In this paper we investigate how to determine the sign of the Gaussian curvature of a timelike minimal surface and determine the behavior of the Gaussian curvature near singular points of a timelike minimal surface.

2 Preliminaries

In this section, we give some notions and known results about timelike minimal surfaces and null curves. We refer [12, 19] for a detailed description of timelike surfaces. The 3-dimensional Lorentz–Minkowski space $\mathbb{R}^3_1$ is the 3-dimensional vector space $\mathbb{R}^3$ with the Lorentzian metric

$$\langle\ ,\ \rangle = -(dx^0)^2 + (dx^1)^2 + (dx^2)^2$$

where (x^0, x^1, x^2) are the canonical coordinates of $\mathbb{R}^3$.

Let Σ be a 2-dimensional manifold. An immersion $f : \Sigma \longrightarrow \mathbb{R}^3_1$ is called *timelike* (resp. *spacelike*) if its *first fundamental form* $\mathrm{I} = f^*\langle\ ,\ \rangle$ is a Lorentzian (resp. Riemannian) metric on Σ. For a timelike immersion f and its unit normal vector field ν, the *shape operator* S and the *second fundamental form* II are defined as follows:

$$df(S(X)) = -\overline{\nabla}_X \nu, \quad \mathrm{II}(X, Y) = \langle \overline{\nabla}_{df(X)} df(Y) - df(\nabla_X Y), \nu\rangle,$$

where X and Y are vector fields on Σ, ∇, and $\overline{\nabla}$ are the Levi–Civita connections on Σ and $\mathbb{R}^3_1$, respectively. The *mean curvature* H and the *Gaussian curvature* K are defined as $H = \mathrm{tr}\, \mathrm{II}/2$ and $K = \det S$, respectively. The eigenvalues of S are called the *principal curvatures* of f. For any surface in $\mathbb{E}^3$ or spacelike surface in $\mathbb{R}^3_1$, the shape operator S is always diagonalizable over $\mathbb{R}$ and we can take real principal curvatures κ_1 and κ_2 of such surface. Therefore, the discriminant of S satisfies

$$H^2 - K = \left(\frac{\kappa_1 + \kappa_2}{2}\right)^2 - \kappa_1\kappa_2 = \left(\frac{\kappa_1 - \kappa_2}{2}\right)^2 \geq 0.$$

In particular, for minimal surfaces in $\mathbb{E}^3$ and maximal surfaces in $\mathbb{R}^3_1$, the Gaussian curvature satisfies $K \leq 0$. We remark that this Gaussian curvature for a spacelike surface is extrinsic, and the intrinsic Gaussian curvature is always nonnegative (see [12]). A point p is called a *umbilic point* if the second fundamental form II is a multiple of the first fundamental form I at p. Flat points of a minimal surface in $\mathbb{E}^3$ or a maximal surface in $\mathbb{R}^3_1$ consist of umbilic points. On the other hand, for a timelike minimal surface, the discriminant of the shape operator $H^2 - K$ can be taken any real value, that is,

(i) S is diagonalizable over $\mathbb{R}$, in this case $H^2 - K \geq 0$ and the equality holds on umbilic points,
(ii) S is diagonalizable over $\mathbb{C}\backslash\mathbb{R}$, in this case $H^2 - K < 0$,
(iii) S is non-diagonalizable over $\mathbb{C}$, in this case $H^2 - K = 0$. A point satisfying this condition is called *quasi-umbilic point* ([4]).

In particular, flat points of a timelike minimal surface consist of umbilic points and quasi-umbilic points.

It is known that at each point of a timelike surface $f : \Sigma \longrightarrow \mathbb{R}^3_1$, there exists a local coordinate system (u, v) on which the first fundamental form I can be written as $\mathrm{I} = 2F du dv$ with a nonzero function F. This local coordinate system is called a *null coordinate system*. A regular curve $\gamma : I \subset \mathbb{R} \longrightarrow \mathbb{R}^3_1$ satisfying $\langle \gamma', \gamma' \rangle = 0$ is called a *null curve*. A null coordinate system is a coordinate system on which the images of coordinate curves are null curves.

On a null coordinate system (u, v), the mean curvature H of a timelike surface f and its unit normal vector field ν satisfy $H\nu = \frac{2}{F}\frac{\partial^2 f}{\partial u \partial v}$. Therefore, we obtain the well-known representation formula by McNertney [13]:

Fact 1 ([13]) *Let $\varphi(u)$ and $\psi(v)$ be null curves in $\mathbb{R}^3_1$ such that $\varphi'(u)$ and $\psi'(v)$ are linearly independent for all u and v. Then*

$$f(u, v) = \frac{\varphi(u) + \psi(v)}{2} \tag{1}$$

is a timelike minimal surface. Conversely any timelike minimal surface can be written as (1) for some two null curves.

In this chapter, we investigate the behavior of the Gaussian curvature of a timelike minimal surface by looking at two null curves which generate the surface.

Example 1 Let us take the null curve $\gamma(t) = (t, \cos t, \sin t)$ and $\varphi(u) = \gamma(u)$, $\psi(v) = \gamma(v)$. The timelike minimal surfaces

$$f_{\pm}(u, v) = \frac{\varphi(u) \pm \psi(v)}{2}$$

are called the *timelike elliptic helicoid* and *timelike elliptic catenoid*, respectively (see, for example, [9]). These surfaces have singular points, that is, points on which the maps $f_{\pm}$ are not immersed (see Fig. 1). The elliptic helicoid f_+ is an inner part of the usual helicoid in $\mathbb{E}^3$.

In this chapter, we consider the following class of timelike minimal surfaces with singular points of rank one, which are called *minfaces* introduced in [16] (see also [1]):

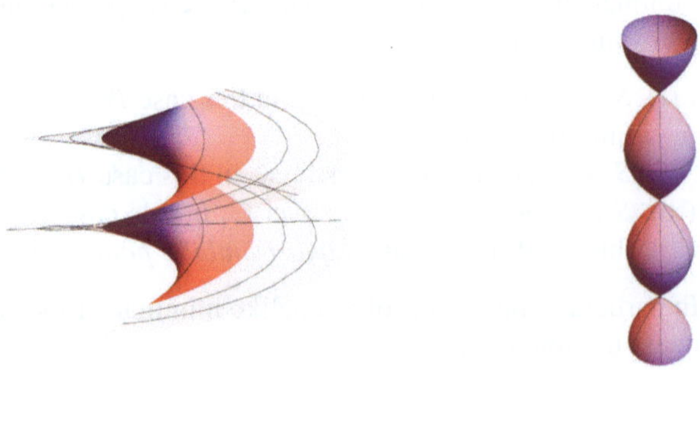

Fig. 1 The elliptic helicoid and catenoid

Definition 1 A smooth map $f : \Sigma \longrightarrow \mathbb{R}^3_1$ is called a *minface* if at each point of Σ there exists a local coordinate system (u, v) in a domain U, functions $g_1 = g_1(u)$, $g_2 = g_2(v)$, and 1-forms $\omega_1 = \hat{\omega}_1(u)du$, $\omega_2 = \hat{\omega}_2(v)dv$ with $g_1(u)g_2(v) \neq 1$ on an open dense set of U and $\hat{\omega}_1 \neq 0$, $\hat{\omega}_2 \neq 0$ at each point on U such that f can be decomposed into two null curves

$$f(u,v) = \frac{1}{2}\int_{u_0}^{u} \left(-1-(g_1)^2, 1-(g_1)^2, 2g_1\right)\omega_1 + \frac{1}{2}\int_{v_0}^{v} \left(1+(g_2)^2, 1-(g_2)^2, -2g_2\right)\omega_2 + f(u_0, v_0). \qquad (2)$$

We denote the above two null curves as $\varphi = \varphi(u)$ and $\psi = \psi(v)$. The quadruple $(g_1, g_2, \omega_1, \omega_2)$ is called the *real Weierstrass data* of f.

A *singular point* of a minface f is a point of Σ on which f is not immersed, and the set of singular points on U of a minface f corresponds to the set $\{(u, v) \in U \mid g_1(u)g_2(v) = 1\}$.

Remark 1 In [16], Takahashi originally gave the notion of minfaces by using the notion of para-Riemann surfaces. To study the local behavior of the Gaussian curvature near singular points of timelike minimal surfaces, we adopt the above definition. For a more detailed exposition on minfaces, see Appendix A in [1].

We investigate the behavior of the Gaussian curvature of timelike minimal surfaces and minfaces from a viewpoint of null curves. In the end of this section, we give notions of non-degeneracy and orientations of null curves.

Definition 2 (*cf.* [5, 15]) A null curve $\gamma : I \to \mathbb{R}^3_1$ is called *non-degenerate* (resp. *degenerate*) at $t \in I$ if γ' and γ'' are linearly independent (resp. dependent) at $t \in I$. A null curve $\gamma : I \to \mathbb{R}^3_1$ is called a *non-degenerate null curve* if γ is non-degenerate at every point.

As pointed out in [15], a null curve is non-degenerate at $t \in I$ if and only if $\det[\gamma'(t)\ \gamma''(t)\ \gamma'''(t)] \neq 0$. Hence, we can define the notion of the orientation for non-degenerate null curves as follows:

Definition 3 For a non-degenerate null curve γ, we define the *orientation* of γ by the sign of the $\det[\gamma'\ \gamma''\ \gamma''']$.

A non-degenerate null curve γ has positive (resp. negative) orientation if γ' moves anticlockwise (resp. clockwise) on the lightcone

$$\mathbb{Q}^2 := \left\{v = (v^0, v^1, v^2) \in \mathbb{R}^3_1 \mid \langle v, v\rangle = 0,\ v^0 \neq 0\right\}$$

as the time coordinate x^0 increases.

3 The Behavior of the Gaussian Curvature Near Regular Points

In Sect. 2 we saw that flat points of a timelike minimal surface consist of umbilic points and quasi-umbilic points. First, we give a characterization of these flat points on the regular part of a minface.

Proposition 1 *Let $p \in \Sigma$ be a regular point of a minface f and φ and ψ be two generating null curves of f. Then the following statements hold.*

(1) p is an umbilic point if and only if both φ and ψ are degenerate at p.
(2) p is a quasi-umbilic point if and only if only one of φ or ψ is degenerate at p.

Therefore, we conclude that flat points can be characterized by the degeneracy of two generating null curves.

Next, we investigate how to determine the sign of the Gaussian curvature of the regular part of a minface f in the Eq. (2). By Proposition 1, we conclude that the two generating null curves φ and ψ in the Eq. (2) are non-degenerate near non-flat points, and hence we can take a null coordinate system (u, v) such that a minface f can be written as

$$f(u,v) = \frac{1}{2}\int_{u_0}^{u}\left(-1-(g_1)^2, 1-(g_1)^2, 2g_1\right)\frac{-\varepsilon_\varphi}{2g_1'}du + \frac{1}{2}\int_{v_0}^{v}\left(1+(g_2)^2, 1-(g_2)^2, -2g_2\right)\frac{-\varepsilon_\psi}{2g_2'}dv + f(u_0, v_0), \qquad (3)$$

where $\varepsilon_\varphi = \pm 1$ and $\varepsilon_\psi = \pm 1$, which represent the orientations of φ and ψ, respectively. Let us consider the local coordinate system $(x, y) = (\frac{u-v}{2}, \frac{u+v}{2})$ associated to the null coordinate system (u, v). Then the first fundamental form I, the second fundamental form II, and the Gaussian curvature K on (x, y) can be computed as

$$\mathrm{I} = \frac{(1-g_1g_2)^2}{4g_1'g_2'}(-dx^2+dy^2), \quad \mathrm{II} = \frac{1}{2}(\varepsilon_1 - \varepsilon_2) + (\varepsilon_1 + \varepsilon_2)dxdy, \quad K = \frac{\varepsilon_1\varepsilon_2}{E^2}.$$

Therefore we have the following characterization of the sign of the Gaussian curvature of a minface:

Theorem 1 *Away from flat points of each minface $f : \Sigma \to \mathbb{R}^3_1$, the sign of the Gaussian curvature K is positive (resp. negative) if and only if the two generating non-degenerate null curves φ and ψ have the same orientation (resp. different orientations). In this case, the local coordinate system $(x, y) = (\frac{u-v}{2}, \frac{u+v}{2})$ associated to the null coordinate system (u, v) in (3) is a conformal asymptotic (resp. conformal curvature line) coordinate system.*

Remark 2 Milnor [14] gave a similar formula which controls the sign of the Gaussian curvature by using Euclidean arclength parameters of null curves. In addition to another proof of Milnor's result, we also give a construction method of conformal asymptotic and conformal curvature line coordinate systems in Theorem 1.

Example 2 The generating null curves of the elliptic helicoid f_+ in Example 1 have the same orientation. By Theorem 1, the sign of the Gaussian curvature of the surface is positive and hence we can take only complex principal curvatures on this surface. On the other hand, the generating null curves of the elliptic catenoid f_- have different orientations. Therefore, the sign of the Gaussian curvature is negative and hence we can take real principal curvatures on this surface.

4 The Behavior of the Gaussian Curvature Near Singular Points

In this section, we investigate the behavior of the Gaussian curvature near singular points on minfaces.

Let U_i be domains of $\mathbb{R}^2$, $p_i \in U_i$ $(i = 1, 2)$. Two smooth maps $f_1 : U_1 \longrightarrow \mathbb{R}^3$ and $f_2 : U_2 \longrightarrow \mathbb{R}^3$ are *$\mathscr{A}$-equivalent* at the points $p_1 \in U_1$ and $p_2 \in U_2$ if there exist local diffeomorphisms Φ of $\mathbb{R}^2$ with $\Phi(p_1) = p_2$ and Ψ of $\mathbb{R}^3$ with $\Psi(f_1(p_1)) = f_2(p_2)$ such that $f_2 = \Psi \circ f_1 \circ \Phi^{-1}$. A *singular point* of a smooth map $f : U \longrightarrow \mathbb{R}^3$ is a point on which f is not immersed, and a singular point is called a *cuspidal edge*, *swallowtail*, or *cuspidal cross cap* if f is $\mathscr{A}$-equivalent to the following maps f_C, f_S, or f_{CCR} at the origin, respectively (see Fig. 2):

$$f_C(u, v) = (u^2, u^3, v), \quad f_S(u, v) = (3u^4 + u^2v, 4u^3 + 2uv, v), \quad f_{CCR}(u, v) = (u, v^2, uv^3).$$

For the spacelike case, Umehara and Yamada [18] introduced the notion of *maxfaces* as a class of maximal surfaces with singular points of rank one, and Fujimori, Saji, Umehara, and Yamada proved in [6] that the singular points of maxfaces in $\mathbb{R}^3_1$ generically consist of cuspidal edges, swallowtails, and cuspidal cross caps. Similarly, these singular points frequently appear on minfaces, and Takahashi [16] gave useful criteria for these singular points by using the real Weierstrass data in Definition 1 as follows:

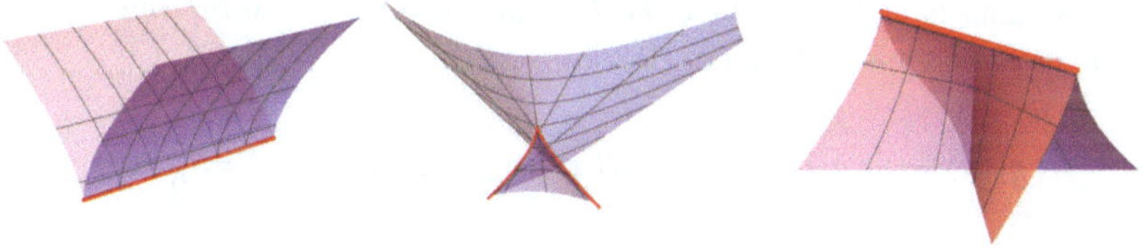

Fig. 2 A cuspidal edge, swallowtail, cuspidal cross cap (where red curves represents the image of singular points)

Fact 2 ([16]) *Let $f : \Sigma \longrightarrow \mathbb{R}^3_1$ be a minface with the real Weierstrass data $(g_1, g_2, \omega_1 = \hat{\omega}_1 du, \omega_2 = \hat{\omega}_2 dv)$ and $p \in \Sigma$ a singular point of f. Then*

(i) p is a cuspidal edge if and only if

$$\frac{g_1'}{g_1^2\hat{\omega}_1} - \frac{g_2'}{g_2^2\hat{\omega}_2} \neq 0 \text{ and } \frac{g_1'}{g_1^2\hat{\omega}_1} + \frac{g_2'}{g_2^2\hat{\omega}_2} \neq 0 \text{ at } p,$$

(ii) p is a swallowtail if and only if

$$\frac{g_1'}{g_1^2\hat{\omega}_1} - \frac{g_2'}{g_2^2\hat{\omega}_2} \neq 0, \ \frac{g_1'}{g_1^2\hat{\omega}_1} + \frac{g_2'}{g_2^2\hat{\omega}_2} = 0, \text{ and } \left(\frac{g_1'}{g_1^2\hat{\omega}_1}\right)' \frac{g_2'}{g_2} - \left(\frac{g_2'}{g_2^2\hat{\omega}_2}\right)' \frac{g_1'}{g_1} \neq 0 \text{ at } p,$$

(iii) p is a cuspidal cross cap if and only if

$$\frac{g_1'}{g_1^2\hat{\omega}_1} - \frac{g_2'}{g_2^2\hat{\omega}_2} = 0, \ \frac{g_1'}{g_1^2\hat{\omega}_1} + \frac{g_2'}{g_2^2\hat{\omega}_2} \neq 0, \text{ and } \left(\frac{g_1'}{g_1^2\hat{\omega}_1}\right)' \frac{g_2'}{g_2} + \left(\frac{g_2'}{g_2^2\hat{\omega}_2}\right)' \frac{g_1'}{g_1} \neq 0 \text{ at } p.$$

By Proposition 1, Theorem 1, and Fact 2, we can prove the following theorem:

Theorem 2 *Let $f : \Sigma \longrightarrow \mathbb{R}^3_1$ be a minface and $p \in \Sigma$ a singular point of f. If p is a*

(i) cuspidal edge, then there are no umbilic points near p,
(ii) swallowtail, then there are umbilic and quasi-umbilic points near p. Moreover, the sign of the Gaussian curvature K is negative near p and $\lim_{q\to p} K(q) = -\infty$,
(iii) cuspidal cross cap, then there are no umbilic and quasi-umbilic points near p. Moreover, the sign of the Gaussian curvature K is positive near p and $\lim_{q\to p} K(q) = \infty$.

By Theorem 2, we can always take real (resp. complex) principal curvatures near swallowtails (resp. cuspidal cross caps) on a minface. Near cuspidal edges, we cannot determine the sign of the Gaussian curvature in general. For example, we can construct minfaces with positive or negative Gaussian curvature along cuspidal edges.

Example 3 (*Enneper surfaces*) The minface constructed from the following real Weierstrass data is called the *timelike Enneper surface (of isothermic type)* or an *analogue of Enneper's surface* [8, 10, 16]:

$$g_1(u) = u, \ g_2(v) = -v, \ \omega_1(u) = \frac{1}{2}du, \ \text{ and } \omega_2(v) = \frac{1}{2}dv.$$

The generating null curves φ and ψ are

$$\varphi(u) = \frac{1}{2}(-u - \frac{u^3}{3}, u - \frac{u^3}{3}, u^2), \quad \psi(v) = \frac{1}{2}(v + \frac{v^3}{3}, v - \frac{v^3}{3}, v^2).$$

Fig. 3 The Enneper's minimal surface in $\mathbb{E}^3$, maximal Enneper surface in ¥$\mathbb{R}^3_1$ and timelike Enneper surface

Moreover, these two null curves φ and ψ are non-degenerate and have negative and positive orientations, respectively. By Theorem 1, the sign of the Gaussian curvature is negative. Unlike the Enneper's minimal surface in $\mathbb{E}^3$, maximal and timelike Enneper surfaces in $\mathbb{R}^3_1$ have singular points. It is known that the maximal Enneper surface in $\mathbb{R}^3_1$ has cuspidal edges, swallowtails, and cuspidal cross caps ([6, 18]). On the other hand, the timelike Enneper surface in $\mathbb{R}^3_1$ has cuspidal edges and two swallowtails (see Fig. 3). Since the sign of the Gaussian curvature is negative, the surface cannot have cuspidal cross caps by Theorem 2.

Remark 3 In Theorem 2, we investigated the behavior of the Gaussian curvature near cuspidal edges, swallowtails, and cuspidal cross caps. Recently in [1], the author has generalized Theorem 2 for any non-degenerate singular point on minfaces.

Acknowledgements The author would like to express his gratitude to Professor Miyuki Koiso for her helpful advices and suggestions. He also thanks the referee for valuable comments. This work was supported by Grant-in-Aid for JSPS Fellows Number 15J06677.

References

1. S. Akamine, *Behavior of the Gaussian Curvature of Timelike Minimal Surfaces with Singularities*, submitted, arXiv:1701.00238
2. G. Bellettini, J. Hoppe, M. Novaga, G. Orlandi, Closure and convexity results for closed relativistic strings. Complex Anal. Oper. Theory **4**, 473–496 (2010)
3. S. Chandrasekhar, Surface tension of liquid crystals. Mol. Cryst. **2**, 71–80 (1966)
4. J.N. Clelland, Totally quasi-umbilic timelike surfaces in $\mathbb{R}^{1,2}$. Asian J. Math. **16**, 189–208 (2012)
5. S. Fujimori, Y.W. Kim, S.-E. Koh, W. Rossman, H. Shin, M. Umehara, K. Yamada, S.-D. Yang, Zero mean curvature surfaces in Lorentz-Minkowski 3-space and 2-dimensional fluid mechanics. Math. J. Okayama Univ. **57**, 173–200 (2015)
6. S. Fujimori, K. Saji, M. Umehara, K. Yamada, Singularities of maximal surfaces. Math. Z. **259**, 827–848 (2008)
7. A. Honda, M. Koiso, Y. Tanaka, Non-convex anisotropic surface energy and zero mean curvature surfaces in the Lorentz-Minkowski space. J. Math-for-Ind. **5**, 73–82 (2013)

8. J. Inoguchi, M. Toda, Timelike minimal surfaces via loop groups. Acta Appl. Math. **63**, 313–355 (2004)
9. Y.W. Kim, S.-E. Koh, H. Shin, S.-D. Yang, Spacelike maximal surfaces, timelike minimal surfaces, and Björling representation formulae. J. Korean Math. Soc. **48**, 1083–1100 (2011)
10. J. Konderak, A Weierstrass representation theorem for Lorentz surfaces. Complex Var. Theory Appl. **50**(5), 319–332 (2005)
11. S. Lee, Weierstrass representation for timelike minimal surfaces in Minkowski 3-space. Commun. Math. Anal., Conf. 01, 11–19 (2008)
12. R. López, Differential Geometry of curves and surfaces in Lorentz-Minkowski space. Int. Electron. J. Geom. **7**, 44–107 (2014)
13. L. McNertney, One-parameter families of surfaces with constant curvature in Lorentz 3-space. Ph.D. thesis, Brown University, 1980
14. T.K. Milnor, Entire timelike minimal surfaces in $\mathbb{E}^{3,1}$. Mich. Math. J. **37**, 163–177 (1990)
15. Z. Olszak, A note about the torsion of null curves in the 3-dimensional Minkowski spacetime and the Schwarzian derivative. Filomat **29**, 553–561 (2015)
16. H. Takahashi, Timelike minimal surfaces with singularities in three-dimensional spacetime. Master thesis, Osaka University, 2012, Japanese
17. J.E. Taylor, Crystalline variational problems. Bull. Amer. Math. Soc. **84**, 568–588 (1978)
18. M. Umehara, K. Yamada, Maximal surfaces with singularities in Minkowski space. Hokkaido Math. J. **35**, 13–40 (2006)
19. T. Weinstein, *An Introduction to Lorentz Surfaces*. De Gruyter Exposition in Mathematics, vol. 22 (Walter de Gruyter, Berlin, 1996)

Accounting for Modelling Errors in Parameter Estimation Problems: The Bayesian Approximation Error Approach

Ruanui Nicholson, Anton Gulley, Jari Kaipio and Jennifer Eccles

Abstract Many parameter estimation problems are highly sensitive to errors. The Bayesian framework provides a methodology for incorporating these errors into our inversion. However, how to characterise the errors in a way that can be efficiently utilised remains a problem in many inversions. Recently the Bayesian approximation error method has been utilised as a systematic way of characterising errors that arise from inaccuracies in the model. We describe the Bayesian approximation error method and demonstrate its use in a homogenisation example. In this example, it is shown that the coarse scale homogenised parameter can be estimated by accounting for the significant modelling error using the Bayesian approximation error method. This modelling error arises from inverting using a model that does not account for the fine scale and has a coarse finite element discretisation.

Keywords Bayesian inversion · Modelling errors · Homogenisation

1 Introduction

Modelling errors arise in nearly all parameter estimation problems. This is because our models are unable to capture the exact real-world physics, or we have deliberately

R. Nicholson · A. Gulley (✉) · J. Kaipio
Department of Mathematics, The University of Auckland,
Private Bag 92019, 1142 Auckland, New Zealand
e-mail: anton.gulley@auckland.ac.nz; gullant@gmail.com

R. Nicholson
e-mail: rnic052@aucklanduni.ac.nz

J. Kaipio
e-mail: j.kaipio@auckland.ac.nz

J. Eccles
School of Environment, The University of Auckland, Private Bag 92019,
1142 Auckland, New Zealand
e-mail: j.eccles@auckland.ac.nz

R. S. Anderssen et al. (eds.), *Agriculture as a Metaphor for Creativity in All Human Endeavors*, Mathematics for Industry 28,
https://doi.org/10.1007/978-981-10-7811-8_6

used surrogate and downscaled models for computational efficiency. The Bayesian approximation error (BAE) method is a technique that has been used within the Bayesian framework to account for various modelling errors. Some recent examples of modelling errors accounted for with the BAE method include: model reduction in full wave ultrasound tomography [2], uncertain boundaries in electrical impedance tomography [3], and approximating the poroelastic wave equation with the elastic wave equation in aquifer parameter estimation [4].

We present the theory of the BAE method and explain how it can be implemented using an example from homogenisation of the Poisson equation.

2 Theory

Let x be some parameter(s) we wish to estimate and let y be measurements which depend on x. The measurements y also include errors, e. We also have the inexact but computationally feasible model $A(x)$ that we wish to use to obtain our estimates such that we can naively write

$$y \approx A(x) + e(x), \tag{1}$$

where we have assumed an additive error model. We also assume that x comes from a prior probability density $\pi_x(x)$ and e also has a probability density $\pi_e(e|x)$ where '$|x$' implies that this may depend on x. A standard Bayesian approach for inverse problems involves using Bayes 璺 · theorem and integrating over all e [1]. This then leaves us with the posterior probability density which is the probability density of x given the data y,

$$\pi(x|y) \propto \pi_e(y - A(x)|x)\pi_x(x). \tag{2}$$

The derivation of Eq. (2) does not take into account the approximate nature of Eq. (1). In many cases the posterior probability density of Eq. (2) will suggest that the true values of x are improbable which is not a good result. This can usually be attributed to the fact that such problems are often ill-posed (in the sense of Hadamard). This means that the estimates of x are highly sensitive to errors, such as the small inaccuracy in Eq. (1), which may have resulted from the inaccuracy of the model $A(x)$.

To improve on these results using the Bayesian approximation error method define an accurate model, A^*. The model A^* may also depend on *nuisance parameters*, x^*. We can now more accurately rewrite Eq. (1) as follows:

$$\begin{aligned} y &= A^*(x|x^*) + e(x), \\ &= A^*(x|x^*) + e(x) + A(x) - A(x), \\ &= A(x) + e(x) + [A^*(x|x^*) - A(x)], \\ &= A(x) + e(x) + \epsilon(x), \end{aligned} \tag{3}$$

where $\epsilon(x) = [A^*(x|x^*) - A(x)]$ is the BAE additive error term. In order for this to be useful, we need to precompute an estimate of a probability density for ϵ. In most cases, it is unfeasible to have an exact model $A^*(x|x^*)$. However, in practice, good results can be achieved by using a third model, $\hat{A}(x|x^*)$. The model $\hat{A}(x|x^*)$ does not need to be a perfect model, it just needs to be *sufficiently* more accurate than A such that the statistics of ϵ are approximately correct [2]. While $\hat{A}(x|x^*)$ is more accurate than $A(x)$, $A(x)$ is still the model that we want to use in the inversion. This may be because $\hat{A}(x|x^*)$ is too computationally intensive or there is no suitable way of using it in a minimisation problem. The model $\hat{A}(x|x^*)$ is only used for offline forward simulations to pre-compute the statistics of $\epsilon(x)$. These statistics are normally estimated by:

1. Randomly generating an ensemble of $\{x^{(i)}, x^{*(i)}\}$ from the prior distribution.
2. Computing an ensemble of $\epsilon^{(i)} = [\hat{A}(x^{(i)}, x^{*(i)}) - A(x^{(i)})]$.
3. Using the statistics of $\epsilon^{(i)}, x^{(i)}$ a probability density $\pi_{\epsilon,x}$ (typically approximated as Gaussian) can be formed.
4. Combining $\pi_{\epsilon,x}$ with π_e to form an estimate of $\pi_{e+\epsilon}(e + \epsilon|x)$.

The posterior probability distribution Eq. (2) can then be rewritten as follows:

$$\pi(x|y) \propto \pi_{e+\epsilon}(y - A(x)|x)\pi_x(x). \tag{4}$$

The BAE method, therefore, provides a way of performing inversions with an inaccurate model $A(x)$. This method has been shown to significantly improve the feasibility of the inversion estimates in many situations [2–4]. The additional computational costs of the method are due to offline precomputations of $A^*(x|x^*) - A(x)$ in order to estimate $\pi_{e+\epsilon}$, however this increase in computational cost is likely to be very small compared to using the more accurate model A^* in the inversions.

3 Homogenisation Example

The BAE has successfully been used in homogenisation [5]. This can be demonstrated by considering the 1D Poisson equation

$$\frac{d}{ds}\left(\kappa \frac{du}{ds}\right) = -1, \text{ with } u(0) = u(1) = 0, \tag{5}$$

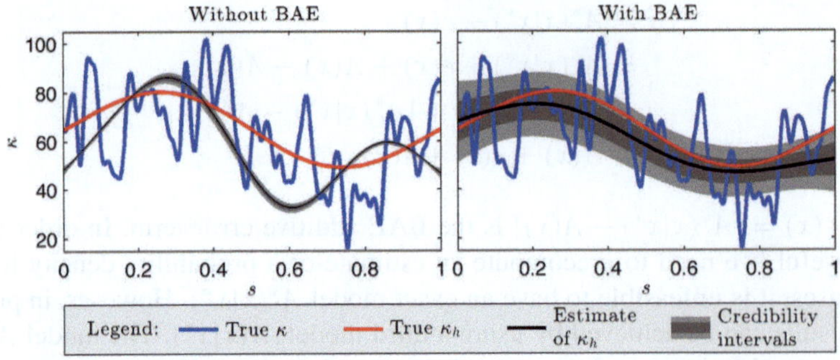

Fig. 1 **Posterior probability densities of** κ_h. The posterior probability densities are approximated to be multivariate normal distributions. The 'Estimate of κ_h' is the approximate posterior mean and the 'Credibility intervals' show the ± 1 and ± 2 standard deviation regions

where $\kappa = \kappa_h + \kappa_m$. The first component, κ_h, is a slowly varying, non-constant, stochastic quantity while κ_m is a rapidly oscillating quantity made up of a spectrum of frequencies. For this homogenisation example, we used the following:

- Synthetic data is generated using $u = A^*(\kappa_h + \kappa_m) + e$, where 5,000 linear finite elements are used to compute A^*. We select 500 evenly spaced measurement points. Randomly generated uncorrelated Gaussian noise, e, is added to the measurements with standard deviation of 0.5% of the root mean square of the signal.
- The approximate model $A(\kappa_h)$ is computed using 500 linear finite elements.
- The more accurate model $\hat{A}(\kappa_h + \kappa_m)$ is computed using 2,000 linear finite elements.
- Correlated Gaussian smoothness priors are used to ensure κ_h and κ_m are restricted to the appropriate bandwidth.

Estimates for the posterior probability densities of k_h are computed with the BAE (Eq. 4) and without the BAE (Eq. 2). These are shown in Fig. 1. It can be seen from Fig. 1 that, in this case, using the BAE does two things: first, it moves the estimates for k_h closer to the true values, and second it significantly increases the (approximate) credibility intervals so that the true k_h is within the ± 2 standard deviation credibility intervals.

References

1. J.P. Kaipio, E. Somersalo, *Statistical and Computational Inverse Problems* (Springer, New York, 2005)
2. J. Koponen, T. Huttunen, T. Tarvainen, J.P. Kaipio, Approximation error method for full-wave tomography. J. Acoust. Soc. Am. **133**, 3230 (2013)
3. D. Calvetti, P.J. Hadwin. J.M.J. Huttunen, J.P. Kaipio, E. Somersalo, Artificial boundary conditions and domain truncation in electrical impedance tomography. Part II: Stochastic extension of the boundary map. Inverse Probl. Imaging **9**(3), 767–789 (2015)

4. T.Lähivaara, N.F. Dudley Ward, T. Huttunen, Z. Rawlinson, J.P. Kaipio, Estimation of aquifer dimensions from passive seismic signals in the presence of material and source uncertainties. Geophys. J. Int. **200**(3), 1662–1675 (2015)
5. R.V. Nicholson, Approaches to multiscale inverse problems. Doctoral thesis, The University of Auckland, Auckland, New Zealand, 2016

Reducing Sludge Formation in the Activated Sludge Process

M. I. Nelson

Abstract The activated sludge process was discovered by Ardern and Lockett in the years 1913– 1914. In the slightly more than 100 years since its discovery, it has become the most widely used process for the biological treatment of both domestic and industrial wastewaters in developed and developing countries. At its most basic, the process consists of an aerated reactor basin connected to a settling unit. The effluent stream leaving the reactor enters the settling unit where particulate matter settles under the action of gravity to the bottom of the unit. From here, it can be recycled into the reactor unit. The recycling of particulate matter is the key to improving the efficiency of the process, as enmeshed within it are micro-organisms. This particulate matter is known as sludge and consequently sludge is good. However, too much sludge is bad; disposal of excess sludge can account for between 50 and 60% of the typical operating costs of the activated sludge process.This chapter provides a historical overview of the activated sludge process and two methods for reducing the amount of sludge: disintegration through the use of a sludge disintegration unit and a biological approach based upon the use of predators that graze upon the sludge.

Keywords Activated sludge process · Predation · Sludge reduction

1 The History of the Activated Sludge Process

In the first half of the nineteenth century, sewage systems emerged as the primary mechanism for the removal of sewage in cities. These replaced older processes which at one extreme simply involved the emptying of chamber pots into the street. How-

M. I. Nelson (✉)
School of Mathematics and Applied Statistics, University of Wollongong,
Wollongong, NSW, Australia
e-mail: mnelson@uow.edu.au

R. S. Anderssen et al. (eds.), *Agriculture as a Metaphor for Creativity in All Human Endeavors*, Mathematics for Industry 28,
https://doi.org/10.1007/978-981-10-7811-8_7

ever, sewage systems do not treat wastewater; they merely move the waste from one location to another. By the mid-1880s, the role played by untreated wastewater in spreading waterborne diseases had become clear. The need to produce a clean, healthy, water supply was exacerbated by both the large-scale production of wastewaters and increasing population densities, both by-products of the burgeoning industrial world. These problems were particularly acute in England, leading to a Royal Commission on River Pollution being established in 1865 and re-established in 1874. The latter lead to the Rivers Pollution Prevention Act (1876).

Some progress treating wastewaters containing pollutants in the form of waste organic matter was made in the second half of the nineteenth century using physicochemical processes and anaerobic methods. Starting in the 1880s, attempts were made at cleaning wastewater using the biological oxidation of the pollutants. This involves bringing together a wastewater containing pollutants with aerobic microorganisms and oxygen. Biological oxidation has the potential to reduce the organic pollutants to a mixture of carbon dioxide, nitrogen, water, and other compounds. These methods had little success because biological oxidation turned out to be a very slow process.

In the years 1913 and 1914, Edward Ardern and William T. Lockett, working at the Davyhulme wastewater treatment plant (Manchester, UK), carried out a series of lab-scale experiments. During these years, they discovered the key step to making aerobic oxidation work. It was already known that aerobic oxidation produced a suspension, or "sludge". In earlier work, this sediment had been removed from the reactor vessel. Ardern and Lockett discovered that if the sediment was retained then the process became significantly more efficient. By retaining the sediment over a series of experiments, they were able to reduce the time for the "full oxidation" of sewage from a period of weeks to one of less than 24 h. Arden and Lockett named the sediment "activated sludge", as it evidently contained an active agent that improved the process, and coined the phrase the "activated sludge process".

At its heart, the activated sludge process essentially entails the use of two units: an aerated biological reactor and a settling unit (or clarifier). In the former, the pollutants are degraded by microorganisms (the active agent that puts the "activated" into "activated sludge"). However, the key to the success of the activated sludge process is the use of a settling unit.

Although not directly realized at the time, aerobic microorganisms flocculate to form settleable solids. These solids are removed from the effluent stream by sedimentation and then returned to the aeration process in a more concentrated culture. It is this recycling of a concentrated activated sludge from the bottom of the clarifier to the biological reactor that drives down the time for "full oxidation" of the wastewater.

Ardern and Lockett presented their findings at a meeting of the Society of Chemical Industry held on the 3rd of April 1914. Following dissemination of their method [1–3] the activated sludge process was rapidly adopted by the wastewater treatment industry. By 1916, during the chaos of the First World War, the first full-scale continuous-flow activated sludge process plant was being used to treat wastewater at Worcester. (Australia's first activated sludge plant, the Glenelg Wastewater Treatment Plant, was fully operational by December 1932 [10]. It was extended in 1941 to include a winery waste treatment stream).

Just over 100 years after their landmark experiments, the activated sludge process is now the most widely used process for the biological treatment of both domestic and industrial wastewaters [26]. For example, of the 375 wastewater treatment plants running in the Netherlands in 2007, 366 (97.6%) were activated sludge systems [22]. Wastewater treatment plants based on the activated sludge process are in widespread use in both developed and developing countries. One of the strengths of this process is its versatility, being used to treat wastewaters from both the domestic (sewage) and industrial sectors. In addition to providing a clean and safe water resource as a consequence of its high efficiency in removing organic pollutants, the process has the attraction of being relatively low in operational costs.

2 Too Much of Anything Is Bad

The degradation of organic pollutants by microorganisms produces new sludge. This fact is central to the success of the activated sludge process—sludge is a self-sustaining resource. Sludge that is produced over the requirements to run the process is known as "excess" sludge. Unfortunately, the activated sludge process is too successful at producing new sludge. The costs associated with the disposal of "excess" sludge can run to 50–60% of the total operating costs of a wastewater treatment plant [4, 5, 16]. Disposal of excess sludge imposes a significant burden on operators as it can be more costly than the wastewater treatment process itself.

In addition to containing beneficial biomass, sludge is a complex mixture of bacteria, heavy metals, inorganic matter (such as phosphorus and nitrogen compounds), organic pollutants, pathogens, and water [21]. The removal of heavy metals from the influent stream into the sludge is often regarded as a side benefit of the activated sludge process [11]. However, as a result of the concentration of heavy metals, and other toxic materials, the disposal of sludge has become increasingly governed by environmental regulations.

The first step in conventional methods for the disposal of excess sludge is dewatering, this converts a water disposal problem to a solid waste disposal problem. In the past, excess dried sludge was commonly disposed of by methods such as incineration, landfilling, dumping at sea, and use as a fertilizer in agriculture. However, any toxic elements in the influent stream of the wastewater treatment plant may become concentrated in the dried sludge. Due to increasing environmental concerns, related to the presence of these toxic elements, older disposal methods are being increasingly regulated. More demanding environmental monitoring inevitably leads to increased operating costs. A second factor impacting operating costs is the decreasing availability of land in urban areas, often associated with increasing population densities. The economic costs of landfilling have, therefore, increased due to a combination of these and other factors [26].

Incineration reduces the volume of solid sludge by upto 95%. Thus this process reduces, but does not eliminate, the demand for landfill sites. Unfortunately, the solid residual from incineration, an ash, contains an increased concentration of noncombustible materials, such as heavy metals and many other toxic compounds. Although the reuse of sludge in the agriculture sector is appealing, because it adds an economic premium to a waste product, the transport costs associated with moving the sludge to the end users is often appreciable.

The combination of the increasing restrictions placed upon the discharge of excess sludge and the associated rising treatment costs has created an impetus to develop methods that reduce the volume and mass of excess sludge. It should be noted that in addition to reducing operating costs sludge reduction has other benefits. For example, sludge reduction can prevent filter beds from being clogged with suspended solids. This maintains their treatment efficiency, consequently this provides a secondary mechanism to reduce operating costs.

In Sects. 2.1 and 2.2, two mechanisms for sludge reduction are discussed. The first of these, Sect. 2.1, is to increase sludge biodegradability by disintegrating it. These techniques, particularly chemical treatments and ozonation, have met with some success. However, the costs associated with running these processes have prevented their wide-spread use. The second of these, Sect. 2.2, is to use predators which grow through consumption of the sludge. This is potentially very attractive, since once the predators have been released into the reactor there are no "running" costs.

2.1 Sludge Reduction Through Sludge Disintegration

The ideal solution to eliminate the problems of the posttreatment disposal of excess sludge is to prevent the excess sludge from being formed in the first place. As sufficient sludge must be produced to maintain the viability of the activated sludge process itself, the aim is to minimize the "excess" sludge. Sludge production can be reduced by a variety of proven techniques including biological, chemical, mechanical, and thermal processes [26]. In general, these methods work by breaking open the cell walls of the bacteria, converting the sludge into a mixture of soluble substrate and particulates.

Amongst a wide variety of techniques, chemical treatments and the use of ozone have become the most widely used in commercial activated sludge plants [17]. Ozone treatments involve moving sludge from the main bioreactor into a separate unit, known as the "sludge disintegrator", where ozone ruptures the cell walls. The treated mixture is then returned to the main bioreactor. Ozonation has been established as a technique that reduces the amount of excess sludge. However, the initial high capital costs and associated ongoing operational costs has restricted its use to niche commercial applications.

2.2 Sludge Reduction Through Predation

A promising alternative to chemical, mechanical, and thermal treatments is sludge reduction through predation by higher organisms, such as protozoa, metazoan, or fungi, upon sludge bacteria [7, 26]. This approach is attractive because it requires little energy and is therefore low cost. Furthermore, unlike chemical treatments this method does not introduce secondary pollutants into the wastewater treatment system [21].

Why does predation reduce sludge? The activated sludge process can be considered to be a food-chain, in which the biomass extracts mass and energy from the substrate. The introduction of a predator introduces a new layer into the food chain: mass and energy are now transferred from the microorganisms to the predator. At each step in the food chain not all of the available energy and material are transferred to the next level: some energy, a significant proportion of the energy, is used for maintenance processes, respiration, and reproduction. Thus predation on microorganisms may lead to a lower total biomass, i.e., sludge reduction.

Some wastewater treatment plants may naturally contain suitable predators. Predators such as metazoan organisms, chiefly Annelida but also including Nematoda and Rotifera, have been found to be present in wastewater treatment plants. It has been suggested that they can enter wastewater treatment plants from surrounding water bodies, or that they can be transported into aeration tanks by birds [15]. Be this as it may, research has focussed on the introduction of 'foreign' predators as a mechanism to control sludge production.

Predation has been shown to be an effective technique in lab-scale experiments [7, 9, 12, 13, 18, 24, 26, 28, 29] and pilot-scale systems [23, 30]. The main thrust in these papers is to quantify the effect that predators have upon process characteristics, in particular to determine the sludge reduction capacity of predators. Although a variety of predators could be used, much attention has focused on the use of worms. Worm growth is clearly a prerequisite for sludge reduction through predation. Relatively little is known about the growth and development of worms during sludge predation. However, it has been shown that the wrong choice of aeration rate, temperature, and predator (worm) density can adversely effect worm growth and consequently sludge reduction [29].

A variety of continuous flow reactor configurations have been used in these investigations. These include: a hydrolyzation food chain reactor [30], a single reactor [12], a single reactor connected to a recycle sludge reactor [9], a membrane bioreactor without biomass discharge [8], a two-stage reactor [7], and a six-stage reactor consisting of alternating aerobic and anaerobic compartments [18].

Despite promising results obtained at lab scale, the role that predictors play in full-scale wastewater treatment plants has rarely been investigated—perhaps because such experiments typically require long-term study carried out over a period of years. The feasibility of using worms to reduce sludge reduction in wastewater treatment plants has been reviewed by Ratsak and Verkuijlen [20]. Experimental investigations at the treatment plant level include [6, 14].

There are two main barriers to the adoption of predation as a cost-effective mechanism to reduce sludge formation. The first of these is the uncontrollable growth of predators [6]. Predator density in full-scale plants can often reach very high densities. Associated with this is a well-known phenomenon in wastewater treatment plants, the so-called "worm blooms", in which predator population densities display peaks followed by a sudden disappearance of the population [19]. The development of methods to control worm proliferation is a challenging problem that needs to be overcome [25]. The second problem is that the use of predators increases the amount of phosphorus, nitrogen, and soluble chemical oxygen demand in both the effluent and waste streams from a treatment plant [26]. This can cause undesirable consequences in receiving waters downstream of treatment plants, i.e., eutrophication and deoxygenation. The release of nutrients and phosphorus into effluent streams is exasperated by predator blooms. Consequently, the release of nutrients and phosphorus due to predation has been investigated [24, 27]. It is essential to know the operating conditions that maintain stable predator populations and which reduce nutrient release.

3 Conclusions

Over the past century, the activated sludge process has emerged to become the most widely used method for the biological treatment of contaminated wastewaters under aerobic conditions. The success of this technique can be ascribed to the use of a settling unit which "captures" particulate matter, allowing it to be recycled into the reactor. This vastly improves the efficiency of the process, since the particulate matter contains entrapped micro-organisms. Consequently, recycling sludge increases the concentration of biomass inside the reactor.

Alas! Too much of a good thing is a bad thing—the disposal of excess sludge imposes a significant overhead on the running of a wastewater treatment plant. Two mechanisms for reducing the amount of sludge produced by the process have been discussed. The first of these is to disintegrate the sludge in situ, converting it into a supply of nutrients. Although this method has been shown to give excellent results on a lab scale, its use in practice are restricted to niche applications. The reason for this is that there are significant costs associated with the use of chemicals and the building and operating of sludge disintegration units. The decrease in costs due to sludge reduction must be balanced by the increase in costs due to the operation of the sludge disintegration unit.

A promising alternative is to introduce predators into the activated sludge plant. The predators grow through consumption of the active biomass, which in turn reduces sludge production. There are conflicting reports in the literature as to whether predation can be successfully implemented as a sludge reduction strategy. If conditions can be found which can guarantee successful operation, then it promises a low-cost route to sludge reduction.

Acknowledgements The author thanks the organizers of the Forum "Math-for-Industry" 2016 for the invitation to give a presentation at the workshop "Agriculture as a metaphor for creativity in all human endeavors" (Queensland University of Technology, Brisbane, Australia; November 21st–23, 2016).

References

1. E. Ardern, W.T. Lockett, Experiments on the oxidation of sewage without the aid of filters. J. Soc. Chem. Ind. **33**, 523–539 (1914)
2. E. Ardern, W.T. Lockett, The oxidation of sewage without the aid of filters. Part II. J. Soc. Chem. Ind. **33**, 122–1124 (1914b)
3. E. Ardern, W.T. Lockett, The oxidation of sewage without the aid of filters. Part III. J. Soc. Chem. Ind. **34**, 937–943 (1915)
4. A.P. Canales, J.L. Poles, Decreased sludge production strategy for domestic wastewater treatment. Water Sci. Technol. **30**, 97–106 (1994)
5. E. Egemen, J. Corpening, N. Nirmalakhandan, Evaluation of an ozonation system for reduced waste sludge generation. Water Sci. Technol. **44**, 445–52 (2001)
6. H.J.H. Elissen, E.T.H.M. Peters, B.R. Buys, A. Klapwijk, Population dynamics of free-swimming Annelida in four Dutch wastewater treatment plants in relation to process characteristics. Hydrobiologia **605**, 131–142 (2008)
7. W. Ghyoot, W. Verstraete, Reduced sludge production in a two-stage membrane-assisted bioreactor. Water Res. **34**, 205–215 (1999)
8. M. Heran, C. Wisniewski, J. Orantes, A. Grasmick, Measurement of kinetic parameters in a submerged aerobic membrane bioreactor fed on acetate and operated without biomass discharge. Biochem. Eng. J. **38**, 70–77 (2008)
9. X. Huang, P. Liang, Y. Qian, Excess sludge reduction induced by Tubifex tubifex in a recycled sludge reactor. J. Biotechnol. **127**, 443–451 (2007)
10. B. Kent, *Special Treatment: Australia's Pioneering Wastewater Plant*. Chemistry in Australia. pp. 24–25 (Dec 2014–Jan 2015)
11. L. Lazzari, P. Sperni, P. Bertin, B. Pavoni, Correlation between inorganic (heavy metals) and organic (PCBs and PAHs) micropollutant concentrations during sewage sludge composting processes. Chemosphere **41**, 427–435 (2000)
12. P. Liang, X. Huang, Y. Qian, Excess sludge reduction in activated sludge process through predation of Aeolosoma hemprichi. Biochem. Eng. J. **28**, 117–122 (2006)
13. P. Liang, X. Huang, Y. Qian, Y. Wei, G. Ding, Determination and comparison of sludge reduction rates caused by microfaunas' predation. Bioresour. Technol. **97**, 854–861 (2006)
14. J. Lou, P. Sun, M. Guo, G. Wu, Y. Song, Simultaneous sludge reduction and nutrient removal (SSRNR) with interaction between Tubificidae and microorganisms: a full-scale study. Bioresour. Technol. **102**, 11132–11136 (2011)
15. G. Milbrink, T. Timm, Distribution and dispersal capacity of the Ponto-Caspian tubificid oligochaete Potamothrix moldaviensis Vejdovský et Mràzek, 1903 in the Baltic Sea Region. Hydrobiologia **463**, 93–102 (2001)
16. O. Nowak, Optimizing the use of sludge treatment facilities at municipal WWTPs. J. Environ. Sci. Health. Part A Toxic/Hazard. Subst. Environ. Eng. **41**, 1807–1817 (2006)
17. Y.-K. Oh, K.-R. Lee, K.-B. Ko, I.-T. Yeom, Effects of chemical sludge disintegration on the performance of wastewater treatment by membrane bioreactor. Water Res. **41**, 2665–2671 (2007)
18. F. Quan, Y. Anfeng, C. Libing, C. Hongzhang, C.-H. Xing, Mechanistic study of on-site sludge reduction in a baffled bioreactor consisting of three series of alternating aerobic and anaerobic compartments. Biochem. Eng. J. **67**, 45–51 (2012)

19. C.H. Ratsak, Effects of Nais elinguis on the performance of an activated sludge plant. Hydrobiologia **463**, 217–222 (2001)
20. C.H. Ratsak, J. Verkuijlen, Sludge reduction by predatory activity of aquatic oligochaetes in wastewater treatment plants: science or fiction? A review. Hydrobiologia **564**, 197–211 (2006)
21. J.H. Rensink, W.H. Rulkens, Using metazoan to reduce sludge production. Water Sci. Technol. **36**, 171–179 (1997)
22. Statistics Netherlands (CBS), 'CBS Statline' webpage. Quoted in [6] (2007), http://statline.cbs.nrl
23. J. Tamis, G. van Schouwenburg, R. Kleerebezem, M.C.M. van Loosdrecht, A full scale work reactor for efficient sludge reduction by predation in a wastewater treatment plant. Water Res. **45**, 5916–5924 (2011)
24. Y. Tian, Y. Lu, Simultaneous nitrification and denitrification process in a new Tubificidae-reactor for minimizing nutrient release during sludge reduction. Water Res. **44**, 6031–6040 (2010)
25. Z. Wang, H. Yu, J. Ma, X. Zheng, Z. Wu, Recent advances in membrane bio-technologies for sludge reduction and treatment. Biotechnol. Adv. **31**, 1187–1199 (2013)
26. Y. Wei, R.T. Van Houten, A.R. Borger, D.H. Eikelboom, Y. Fan, Minimization of excess sludge production for biological wastewater treatment. Water Res. **37**, 4453–4467 (2003)
27. Y. Wei, H. Zhu, Y. Wang, J. Li, P. Zhang, J. Hu, J. Liu, Nutrients release and phosphorus distribution during oligochaetes predation on activated sludge. Biochem. Eng. J. **43**, 239–245 (2009)
28. Q. Zhang, S. Wang, W. Wang, P. Bao, B. Li, Y. Peng, Achieving one-stage sludge reduction by adding Chironomid larvae in wastewater treatment systems. Ecol. Eng. **83**, 291–295 (2015)
29. X. Zhang, Y. Tian, Q. Wang, H. Liu, Waste sludge reduction using Limnodrilus hoffmeisteri: growth, development and sludge predation potential of aquatic worm correlate with process conditions. Ecol. Eng. **58**, 406–413 (2013)
30. T. Zheng, P. Li, Q. Wang, X. Li, H. Ai, K. Gan, A. Sharavsambuu, Pilot-scale experiments on brewery wastewater treatment and sludge reduction based on food chain predation. Desalin. Water Treat. **55**, 1142–1151 (2015)

Incorporating Prior Knowledge in the Calibration of Hydrological Models for Water Resources Forecasting

Julien Lerat

Abstract The management of water resources in Australia faces increasing challenges due the rise of conflicting demands and a highly variable climate. In this context, the Bureau of Meteorology developed a dynamic seasonal forecasting service providing probabilistic forecasts of river flow at selected locations across Australia by coupling rainfall forecasts from a Global Circulation Model with a rainfall–runoff model. The chapter presents a method to improve the Bayesian inference of the rainfall–runoff model parameters by using an informative prior derived from the calibration of the model on a large sample of catchments. This prior is compared with a uniform prior that is currently used in the system. The results indicate that the choice of the prior can have a significant impact on forecast performance for both daily and monthly time steps. The use of an informative prior generally improved the performance, especially for one test catchment at daily time step where prediction intervals were narrowed without compromising forecast reliability. For other catchments and time steps, the improvement was more limited.

Keywords Seasonal streamflow forecasts · Rainfall–runoff modelling
Bayesian inference · Prior distribution · Importance sampling

1 Introduction

The management of water resources in Australia faces increasing challenges due the rise of conflicting demands and a highly variable climate oscillating between extreme droughts and floods. In this context, the prediction of inflows into major reservoirs over the coming months constitutes a critical decision variable for reservoir operations and water allocations. To address this issue, the Bureau of Meteorology, in collaboration with the Commonweath Scientific and Industrial Research Organisation (CSIRO) [18], developed a seasonal forecasting service providing probabilistic

J. Lerat (✉)
Bureau of Meteorology, GPO Box 2334, Canberra, ACT 2601, Australia
e-mail: julien.lerat@bom.gov.au

R. S. Anderssen et al. (eds.), *Agriculture as a Metaphor for Creativity in All Human Endeavors*, Mathematics for Industry 28,
https://doi.org/10.1007/978-981-10-7811-8_8

forecasts of accumulated river flow for the coming 3-month period at selected locations across Australia [3]. Following the initial release of the service in 2010, stakeholders requested two important upgrades: (1) improved synchronisation between seasonal streamflow forecasts and the seasonal climate outlook produced by the Bureau [2], and (2) forecasts issued at monthly time steps. Following these requests, the Bureau initiated the development of a new forecasting system, referred to as the dynamic system [17]. This system couples rainfall forecasts from the Bureau climate outlook model (Predictive Ocean and Atmosphere Model of Australia, or POAMA) with the GR4J rainfall–runoff model [13]. This chapter presents a method to improve the inference of the rainfall–runoff model parameters by using an informative prior.

Figure 1 presents a schematic of the dynamic system including the extraction of climate forecasts from the POAMA climate model, the calibration of the rainfall–runoff model, the generation of runoff forecasts from the rainfall–runoff model and the post-processing of these forecasts to remove residual biases. These components vary greatly in nature, time and space resolution. Nonetheless, it is possible to describe the forecasting process using the following common Bayesian framework [9]:

$$q_C \sim f\left(\bullet \middle| \hat{q}(\theta, I_C), \nu\right) \tag{1}$$

$$(\theta, \nu) \sim g(\bullet) \tag{2}$$

where q_C is a set of observations (e.g., observed runoff data) during the calibration period C, $\hat{q}$ is the output of the GR4J rainfall–runoff model driven by climate inputs I_C over the calibration period, θ are the four GR4J model parameters [13], f is a probability density with parameter vector ν, and g is the prior probability density for θ and ν. To speed-up the inference process, the original GR4J parameters are transformed as follows:

$$\theta = \{\log(X_1), \text{asinh}(X_2), \log(X_3), \log(X_4 - 0.49)\}, \tag{3}$$

where $\{X_k\}_{k=1,..,4}$ are the original parameters of the GR4J model. In the rest of the paper, the parameters θ are simply referred to as "rainfall–runoff parameters". Using this framework, the distribution of θ and ν is first inferred from a set of observed streamflow data $\{q_C(i)\}_{i=1,...,n}$. The joint posterior distribution of θ and ν is given by

$$P(\theta, \nu | q_C) \propto f\left(q_C \middle| \hat{q}(\theta, I_C), \nu\right) g(\theta, \nu). \tag{4}$$

Subsequently, this model is applied in forecast mode to generate a forecast prediction intervals by sampling from the following distribution:

$$P(q_F^* | q_C) = \int_{\theta,\nu,I_F^*} f\left(q_F^* \middle| \hat{q}(\theta, I_F^*), \nu\right) P(\theta, \nu | q_C)\, P(I_F^*)\, d\theta\, d\nu\, dI_F^*, \tag{5}$$

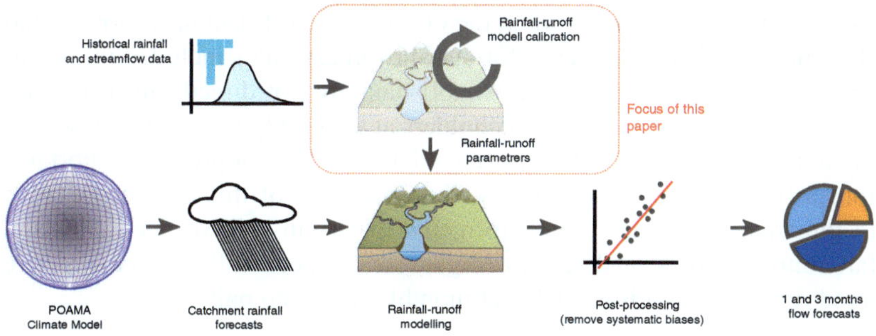

Fig. 1 Overview of the dynamic forecasting system

where I_F^* are the climate inputs during the forecast period (e.g., rainfall forecast) with associated probability $P(I_F^*)$, and q_F^* is the runoff forecast.

Due to the complexity of the dynamic system and the legacy from preexisting components, the framework described in Eqs. 1 to 5 is currently applied to the calibration of the rainfall–runoff model only (red box in Fig. 1). The integration of all components in a Bayesian framework is the topic of several research projects supported by the Bureau and lead by the University of Adelaide and CSIRO within the Water Information Research and Development Alliance (WIRADA). When focusing on the calibration of the rainfall–runoff component, rainfall forecasts are replaced by historical observations. Consequently, the probability model for the climate inputs I_F used in Eq. 5 can be simplified as follows:

$$P(I_F^*) = \prod_i \delta_{I_F(i)}\Big(I_F^*(i)\Big) \tag{6}$$

where $I_F(i)$ and $I_F^*(i)$ are observed and forecast rainfall on day i and δ denote the Dirac delta. As a result, the forecast generation process detailed in Eq. 5 becomes as follows:

$$P(q_F^*|q_C) = \int_{\theta,\nu} f\Big(q_F^*\Big|\hat{q}(\theta, I_F), \nu\Big)\ P(\theta, \nu|q_C)\ d\theta\ d\nu \tag{7}$$

As can be seen in Eqs. 4 and 7, the likelihood f and the prior g are two central components of a Bayesian forecasting system. So far, most of the published work has focused on the definition of a likelihood function that can accurately reproduce the joint distribution of observed and simulated runoff [14, 16]. Few studies have tried to introduce an informative prior on rainfall–runoff parameters. In these attempts, the prior took a simple analytical form like a uniform [6], Beta or Weibull distribution [1]. It will be shown in Sect. 2.2 that these assumptions can be difficult to implement when dealing with a large number of catchments across a wide range of hydroclimatic conditions.

Further complication arises when the observed runoff data q_C is not available at a daily time step. This is the case for certain forecast sites included in the Bureau seasonal forecasting service, where runoff data are not directly measured from a river station but provided by the stakeholders at a monthly time step. These sites are usually associated with the inflow to large dams, and are of primary importance for the forecasting service. Consequently, the objectives of this chapter are twofold: (1) to illustrate the value of an informative priors in a Bayesian calibration of rainfall–runoff models, and (2) investigate more specifically the case where the observed runoff data q_C is available at monthly time step only.

2 Method

In the rest of the chapter, the runoff data q_C and climate inputs I_C over the calibration period C are simply referred to as q and I, respectively.

2.1 *Likelihood Function*

The definition of the likelihood function follows the approach suggested by [12] in which the observed and simulated runoff data are transformed using a Box–Cox transform. This leads to

$$y_i = \frac{q(i)^\lambda - 1}{\lambda} \tag{8}$$

$$\hat{y}_i = \frac{\hat{q}(I, \theta)(i)^\lambda - 1}{\lambda}, \tag{9}$$

where $q(i)$ and $\hat{q}(I, \theta)(i)$ are the observed and simulated runoff data for time step i and λ is the Box–Cox exponent. After data transformation, it is assumed that the model residual follows a centred normal distribution with

$$f(\hat{y}_i - y_i) = \mathcal{N}(\hat{y}_i - y_i \big| \sigma), \tag{10}$$

where $y_i - \hat{y}_i$ is the residual for day i and $\mathcal{N}(\cdot\,|\sigma)$ is the normal probability density function with scale parameter σ. Combining Eqs. 8 to 10 leads to the following expression of the likelihood function:

$$f(q|\hat{q}, \nu) = \prod_i \mathcal{N}(\hat{y}_i - y_i \big| \sigma) \prod_i \frac{\partial y_i}{\partial q_i}, \tag{11}$$

where $\nu = \{\lambda, \sigma\}$ are the residual error parameters. The second term in the right-hand side of Eq. 11 is the Jacobian of the data transform. Following [12], the parameter σ is re-parameterised as

$$\sigma = \sigma_0 \, \bar{q}^{\lambda} \tag{12}$$

where $\bar{q}$ is the average flow during the calibration period. This re-parameterisation facilitates the comparison of parameter values between different sites. Eventually, the error model parameters vector becomes $\{\lambda, \sigma_0\}$.

In the case where the observed runoff data q is monthly, the rainfall–runoff model is run at a daily timestep first, then the simulated runoff is aggregated to monthly and the likelihood is computed from monthly observed and simulated data using Eq. 11.

This configuration creates a difficulty for the inference of the parameters describing the daily dynamic of the rainfall–runoff relationship. More specifically, the fourth parameter of GR4J (θ_4) controls a convolution kernel applied to daily rainfall in order to smooth and delay the simulated runoff. As a result, the likelihood function alone does not constrain this parameter tightly and the role of the prior becomes important as indicated in the following section.

2.2 Priors and Sampling Strategy

To simplify its expression, the prior distribution g of rainfall–runoff and error model parameters mentioned in Eq. 4 is written as follows:

$$g(\theta, \nu) = h_1(\theta) h_2(\sigma_0) h_3(\lambda), \tag{13}$$

where the h_k are the prior distributions for the different parameters. In all the configurations tested in this chapter, h_2 and h_3 are uniform density function defined as

$$h_2(\sigma_0) = \frac{1}{3} \chi_{[0,3]}(\sigma_0) \tag{14}$$

$$h_3(\lambda) = \chi_{[0,1]}(\lambda). \tag{15}$$

wherer χ_U is the indicator function for the interval U.

The main focus of this paper is to compare two alternatives for the prior of the rainfall–runoff model parameters θ. The first prior is a simple uniform distribution over an acceptable parameter space defined as follows:

$$h_1^{UNIF}(\theta) = \chi_{[0,9.2]}(\theta_1) \; \chi_{[-4.6,4.6]}(\theta_2) \; \chi_{[0,8.5]}(\theta_3) \; \chi_{[-4.6,3.7]}(\theta_4). \tag{16}$$

With this prior, the parameters are considered to be uncorrelated and uniformly distributed across the feasible intervals. This type of prior is currently used in the dynamic seasonal forecasting system [11].

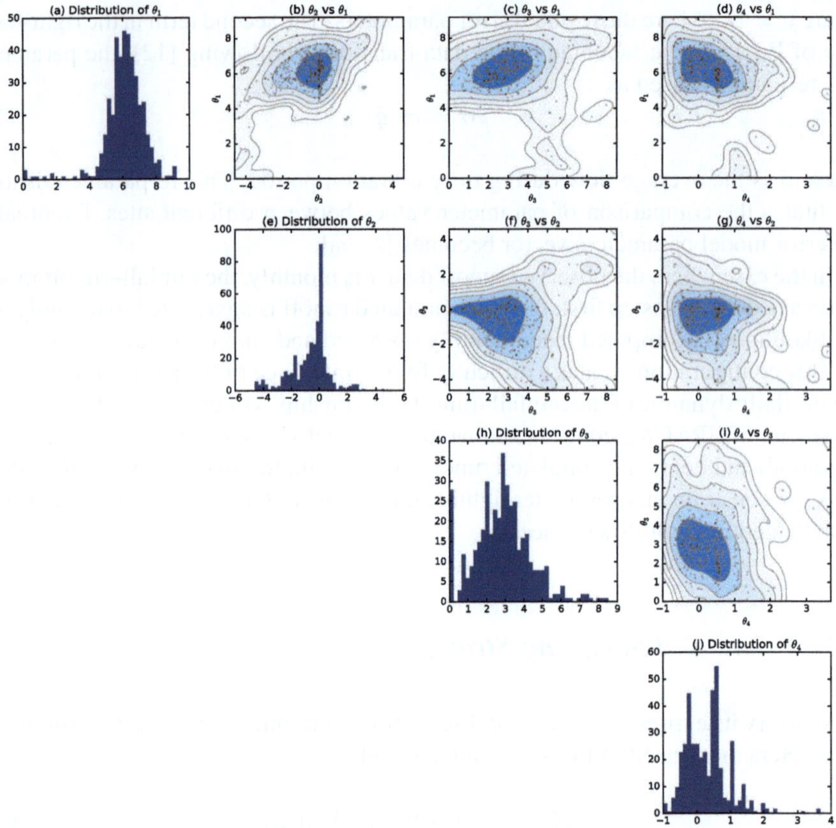

Fig. 2 Distribution of the four GR4J parameter $\theta_1, \theta_2, \theta_3$, and θ_4 across 419 catchments in Australia. A 2D Kernel Density Estimator is used to draw the density contours of the bi-variate distributions

The second prior is built from a large calibration exercise conducted by the Bureau of Meteorology over 419 catchments across Australia. In this exercise, the GR4J model was calibrated by maximising the posterior expressed in Eq. 4 and using the uniform priors described in Eqs. 14 to 16. The resulting distribution of the 419 parameter sets is presented in Fig. 2. We can see that the joint parameter distribution is complex due to multi-modality (e.g., in Fig. 2j), clustering along the parameter bounds (e.g., in Fig. 2h), long tails (e.g., in Fig. 2e) and strong departure from normality in bi-variate distributions (e.g., in Fig. 2f).

Gaussian mixtures were initially trialled, but could not accurately represent the key features visible in Fig. 2 with a parsimonious model. Consequently, such a prior cannot be directly used in a Markov Chain Monte Carlo sampler, which requires a parametric representation of the prior to implement the sampling rule. To overcome this problem, the parameters were inferred using an importance sampling method [15] as follows:

1. 100 error model parameters $\nu = \{\sigma_0, \lambda\}$ were drawn from the uniform prior of Eqs. 14 and 15,
2. the posterior expressed in Eq. 4 is computed for each parameter ν and each one of the 419 parameter sets θ from the informative prior,
3. for each rainfall–runoff parameter set θ_i (with $i \in [1, .., 419]$) and error model parameter set ν_j (with $j \in [1, .., 100]$), the importance weight is computed as

$$\omega_{i,j} = \frac{P(\theta_i, \nu_j | q)}{\sum_{i,j} P(\theta_i, \nu_j | q)}, \tag{17}$$

where P is the posterior distribution of Eq. 4.
4. Finally, the maximum weight ω^* is identified and a set of N parameter sets is retained with weights higher than $10^{-5} \times \omega^*$.

The previous algorithm describes the sampling process using an informative prior. With the uniform prior, the procedure is identical, except that the 419 parameter sets are replaced with a set of 2000 parameters sampled from the uniform distribution presented in Eq. 16.

For both priors, the predictive distribution expressed in Eq. 7 is generated as a set of M =10,000 time series referred to as "ensembles". The ensemble generation process operates as follows. Rainfall–runoff and error model parameters are selected randomly with weighting ω^*, and for each selected parameter set $\{\theta_i, \nu_j\}$:

1. the rainfall–runoff model is run first, producing a simulated flow time series $\hat{q}(\theta_i)$,
2. a set of $M \times \omega$ samples are subsequently drawn from the error model using Eqs. 8 to 10 and parameters ν_j.

2.3 Validation Strategy

The inference process is conducted using data from 1975 to 1995. The simulations are subsequently generated from 1995 to 2010. During this last period, referred to as "validation period", the three following performance metrics are computed:

- the bias skill score is computed as

$$B = 1 - \left| \frac{\sum_i Median(\hat{q}(i))}{\sum_i q(i)} \right|, \tag{18}$$

where $Median(\hat{q}(i))$ is the median of the 10,000 simulated ensembles on day i.
- the Nash–Sutcliffe efficiency is given by

$$NSE = 1 - \frac{\sum_i (Median(\hat{q}(i)) - q(i))^2}{\sum_i (\bar{q} - q(i))^2} \tag{19}$$

- the Continuous Rank Probability Skill Score [5] is computed as

$$CRPSS = 1 - \frac{CRPS(q, \hat{q})}{CRPS(q, q)}, \tag{20}$$

where the score $CRPS$ is given by

$$CRPS(q, \hat{q}) = \sum_i \int_0^{+\infty} [F_{\hat{q}(i)}(x) - H_{q(i)}(x)]^2 dx, \tag{21}$$

where $F_{\hat{q}(i)}$ is the cumulative density function of the simulated flow on day i and H is the Heaveside function. In Eq. 20, the expression $CRPS(q, q)$ corresponds to the CRPS value obtained with a climatological forecast.

The three scores vary between $-\infty$ and 1, with a value of 1 corresponding to a perfect forecast. The value of 0 corresponds to a forecast having a similar performance than a climatological benchmark. The bias and NSE scores focus on the performance of the median forecast, whereas the CRPS measures the performance for the whole forecast ensemble.

A fourth measure of performance is used to check if the prediction intervals correctly bracket the corresponding observations. This aspect of forecast performance is referred to as "forecast reliability" [10] and is quantified with the Probability Integral Transform (PIT). The PIT is defined for each point in the time series as

$$PIT(i) = F_{\hat{q}(i)}(q(i)). \tag{22}$$

If the forecast is reliable, the PIT should be uniformly distributed between 0 and 1. This is checked by plotting the sorted PIT values and comparing the resulting curve with the 1:1 line.

2.4 Test Catchments

The calibration procedure outlined in the previous paragraph was tested on three catchments presented in Fig. 3. The catchments cover a wide range of climatic and hydrologic conditions. The rainfall data were extracted from the Australian Water Availability Project (AWAP) data maintained by the Bureau of Meteorology. The potential evapotranspiration data were computed with the Penman method from AWAP data. Runoff data were extracted from the Water Data Online archive maintained by the Bureau of Meteorology in collaboration with States jurisdictions.

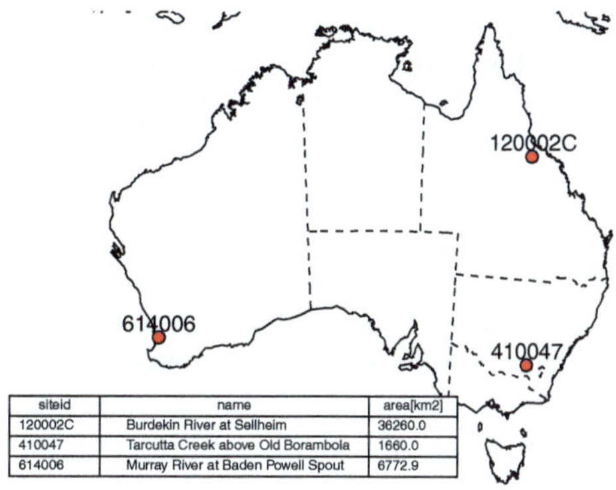

siteid	name	area[km2]
120002C	Burdekin River at Sellheim	36260.0
410047	Tarcutta Creek above Old Borambola	1660.0
614006	Murray River at Baden Powell Spout	6772.9

Fig. 3 Location and characteristics of the three test catchments

3 Results

Figure 4 shows the values of the three performance metrics computed over the validation period for the three test catchments and two inference schemes with uniform (UNIF) and informative (INF) priors. Figures 4a–f present the performance computed from daily and monthly times series, respectively. Among all results displayed in Fig. 4, the use of an informative prior always improve the performance compared to a uniform prior, except for the bias in Tarcutta Creek at daily time step (see Fig. 4a) and the CRPS in Burdekin River at monthly time step (see Fig. 4e). The improvement in performance is significant for the Murray River at daily time step

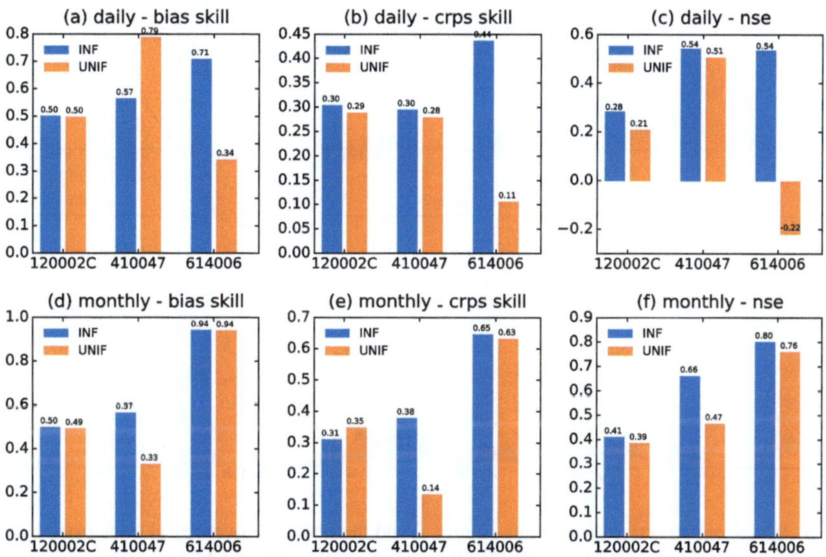

Fig. 4 Performance metrics computed from the validation period for the three test catchments at a daily time step (figures **a**, **b**, and **c**) and monthly time step (figures **d**, **e**, and **f**)

and Tarcutta Creek at monthly time step where all metrics increase by more than 0.1. This improvement is particularly large for the Murray River at daily time step where bias, crps, and nse skill improve from 0.34 to 0.71, 0.11 to 0.44, and −0.22 to 0.54, respectively. These results suggest that the prior plays an important role in the inference scheme and that significant performance gains can be obtained by using an informative prior.

The performance metrics presented in Fig. 4 are an aggregated measure of performance over the validation period and can hide detailed features of the simulation. Figure 5 explores the results obtained in the Murray River further by showing two predictive distributions at daily time step during the largest flood that occurred in the validation period. The distributions are generated with a uniform and informative prior in Fig. 5a top and bottom, respectively. The two distributions differ significantly during the first two flood peaks in July and August 1996, where the distribution generated from the uniform prior is much wider than the one obtained with an informative prior. The opposite occurs during the second flood peak in October 1996, but to a lesser extent. Overall, the use of an informative prior leads to narrower prediction intervals and reduces the prediction bias during the first two peaks. This analysis confirms the significant performance gain obtained with an informative prior highlighted in Fig. 4a–c for this catchment.

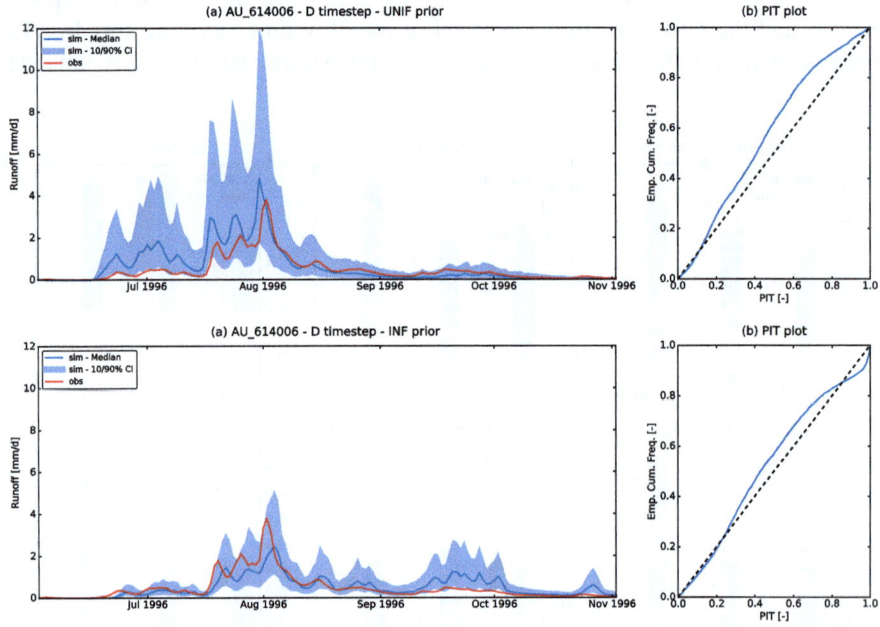

Fig. 5 Posterior predictive distribution (figure a) and their probability integral transform (figure b) for daily simulations in the Murray River (Western Australia, site id 614006) using uniform prior (top) and informative prior (bottom). The observed runoff is shown in red and the simulations in blue in figure a

Figure 5b top and bottom investigate another aspect of the predictive distribution by displaying the distribution of the Probability Integral Transform (PIT) values over the validation period. The two plots suggest that the predictive intervals lack reliability with important deviations from the 1:1 line. More specifically, a large number of low PIT values is present in both predictive distributions, which indicates a tendency to overestimate the observed flow. However, in spite of these limitations, it can be observed that the PIT values corresponding to the informative prior (Fig. 5b bottom) are closer to the 1:1 line compared to the ones produced with the uniform prior (Fig. 5b top). This improvement in reliability combined with the reduction of the prediction interval width described in the previous paragraph constitutes a significant improvement of overall forecast performance.

4 Discussion and Conclusion

This chapter presented a method to improve the calibration of rainfall-runoff model as part of the dynamic seasonal streamflow forecasting system developed by the Bureau of Meteorology. The chapter focused on the choice of the prior distribution and compared a uniform prior with an informative prior derived from the calibration of the model on a large sample of catchments.

The results presented in the previous section indicate that the choice of the prior can have a significant impact on forecast performance for both daily and monthly time steps. The use of an informative prior generally improved the performance, especially for the Murray River at the daily time step where prediction intervals were narrowed without compromising forecast reliability. For other catchments and time steps, the improvement was more limited.

These results are promising, but several points related to the inference scheme will require further developments. First, it was noted that, although the performance in the Murray River improved when comparing a uniform and informative prior, both forecasts lacked reliability. This can be attributed to the limitations of the error model which cannot describe the joint probability of observed and simulated runoff accurately. More specifically, the error model used in this paper did not account for auto-correlation in the residuals. Initial attempts to incorporate this element (not shown) lead to inferior performance, confirming the difficulties encountered by [4] using a similar inference scheme.

Second, the sampling scheme was limited to an importance re-sampling approach where the parameter samples are exclusively determined by the prior distribution. Other approaches developed recently, such as the sequential Monte Carlo sampling described by [7], could be used to perturb the prior samples and avoid a reduction of the sample diversity.

Finally, the study presented in this chapter will be extended to a larger set of test catchments to assess the significance of the trends presented here.

References

1. B.C. Bates, E.P. Campbell, A Markov chain Monte Carlo scheme for parameter estimation and inference in conceptual rainfall-runoff modeling. Water Resour. Res. **37**(4), 937–947 (2001)
2. Bureau of Meteorology, Seasonal Climate Outlook (2016a), http://bom.gov.au/climate/outlook. Accessed 20 Sep 2016
3. Bureau of Meteorology, Seasonal Streamflow Forecasts (2016b), http://bom.gov.au/water/ssf. Accessed 20 Sep 2016
4. G. Evin, D. Kavetski, M. Thyer, G. Kuczera, Pitfalls and improvements in the joint inference of heteroscedasticity and autocorrelation in hydrological model calibration. Water Resour. Res. **49**(7), 4518–4524 (2013)
5. H. Hersbach, Decomposition of the continuous ranked probability score for ensemble prediction systems. Weather Forecast. **15**(5), 559–570 (2000)
6. D. Huard, A. Mailhot, Calibration of hydrological model GR2M using Bayesian uncertainty analysis. Water Resour. Res. **44**(2) (2008)
7. E. Jeremiah, S.A. Sisson, A. Sharma, L. Marshall, Efficient hydrological model parameter optimization with Sequential Monte Carlo sampling. Environ. Model. Softw. **38**, 283–295 (2012)
8. D. Kavetski, G. Kuczera, S.W. Franks, Bayesian analysis of input uncertainty in hydrological modeling: 2. application. Water Resour. Res. **42**(3) (2006)
9. G. Kuczera, D. Kavetski, S. Franks, M. Thyer, Towards a bayesian total error analysis of conceptual rainfall-runoff models: characterising model error using storm-dependent parameters. J. Hydrol. **331**(1), 161–177 (2006)
10. F. Laio, S. Tamea, Verification tools for probabilistic forecasts of continuous hydrological variables. Hydrol. Earth Syst. Sci. **11**(4), 1267–1277 (2007)
11. J. Lerat, C. Pickett-Heaps, D. Shin, S. Zhou, P. Feikema, U. Khan, R. Laugesen, N. Tuteja, G. Kuczera, M. Thyer, D. Kavetski, Dynamic streamflow forecasts within an uncertainty framework for 100 catchments in Australia, in *36th Hydrology and Water Resources Symposium: The Art and Science of Water, Engineers Australia* (2015), p. 1396
12. D. McInerney, M. Thyer, D. Kavetski, G. Kuczera, J. Lerat, Evaluation of approaches for modelling heteroscedasticity in the residual errors of hydrological predictions. Water Resour. Res. accepted (2017)
13. C. Perrin, C. Michel, V. Andréassian, Improvement of a parsimonious model for streamflow simulation. J. Hydrol. **279**(1), 275–289 (2003)
14. G. Schoups, J.A. Vrugt, A formal likelihood function for parameter and predictive inference of hydrologic models with correlated, heteroscedastic, and non-gaussian errors. Water Resour. Res. **46**(10) (2010)
15. A.F. Smith, A.E. Gelfand, Bayesian statistics without tears: a sampling-resampling perspective. Am. Stat. **46**(2), 84–88 (1992)
16. T. Smith, A. Sharma, L. Marshall, R. Mehrotra, S. Sisson, Development of a formal likelihood function for improved bayesian inference of ephemeral catchments. Water Resour. Res. **46**(12) (2010)
17. N. Tuteja, D. Shin, R. Laugesen, U. Khan, Q. Shao, E. Wang, M. Li, H. Zheng, G. Kuczera, D. Kavetski, G. Evin, Experimental evaluation of the dynamic seasonal streamflow forecasting approach. Technical Report (Australian Bureau of Meteorology, 2012)
18. Q. Wang, D.L. Shrestha, D. Robertson, P. Pokhrel, A log-sinh transformation for data normalization and variance stabilization. Water Resour. Res. **48**(5) (2012)

Maintaining Reliable Agriculture Productivity and Goyder's Line of Reliable Rainfall

Julia Piantadosi and Robert S. Anderssen

Abstract Our aim in this study is to generate rainfall totals using multidimensional copulas designed to simulate realistic rainfall statistics that inform analysis of current rainfall patterns and enables better projections for a comprehensive range of future scenarios which can be used as input to ecological models including yield crop simulations for management and risk assessment. To demonstrate the mathematical models, we consider a Case Study of Goyder's line of *reliable rainfall* and the goal of maintaining reliable agriculture productivity in South Australia. We will present the results from the rainfall models using copulas of maximum entropy and discuss how they can be used to assist with management of land use in South Australia.

Keywords Copulas · Rainfall · Reliable agriculture productivity

1 Introduction

Australia is blessed with great soils at various locations around the country such as in the southern region of South Australia. However, maintaining reliable agriculture productivity agriculture requires water. The three possibilities are above surface irrigation, aquifer irrigation and rainfall. In the southern region of South Australia, rainfall is the main source using stored and desalinated water. Generated rainfall totals at various timescales are commonly used as input for simulation of hydrological and ecological systems to model a wide range of scenarios to assist decision-making and risk assessment. In rainfall modelling, much has been made of recent fluctuations in annual rainfall. Previous work on rainfall simulation has already shown that variation in annual rainfall over a period of 150 years could be much larger than previously

J. Piantadosi (✉)
University of South Australia, Adelaide, Australia
e-mail: juliapiantadosi@bigpond.com

R. S. Anderssen
CSIRO Data61, Canberra, Australia
e-mail: Bob.Anderssen@data61.csiro.au

R. S. Anderssen et al. (eds.), *Agriculture as a Metaphor for Creativity in All Human Endeavors*, Mathematics for Industry 28,
https://doi.org/10.1007/978-981-10-7811-8_9

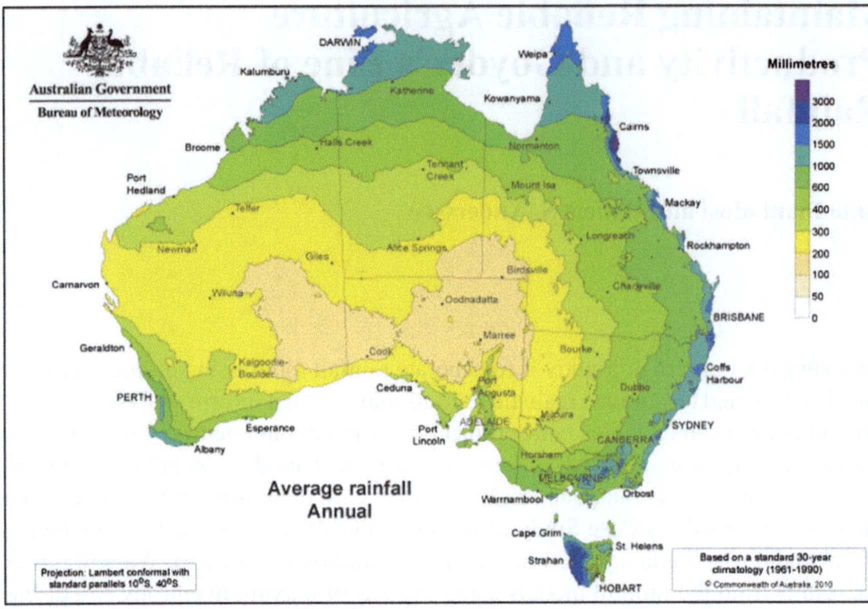

Fig. 1 Annual rainfall map of Australia based on the standard 30 years period 1961 to 1990

thought and indeed much larger than the observed variation [2, 11]. Do farmers in dry countries such as Australia need to be more cautious in their reliance on rainfall records even if they ignore the uncertainties of climate change? Are prolonged periods of low or high rainfall more common than our limited records would suggest? The proposed work will help us answer these difficult questions and give us a greater understanding of the likely impact of fluctuations in rainfall [2, 11–13]. This leads naturally to the need to identify the regions in Australia for which reliable agriculture productivity is a possibility. Historically, an early attempt to do that is Goyder's line of *reliable rainfall* in South Australia.

2 Rainfall

Water is a critical resource, and concern about climate change has generated recent interest in rainfall modelling to enable improved water management practices. This is of critical importance to Australia. The annual rainfall map of Australia shows that extensive areas in central Australia are generally very dry (Fig. 1), with average annual rainfall below 600 millimetres (mm) over 80% of the continent and below 300 mm over 50% [3].

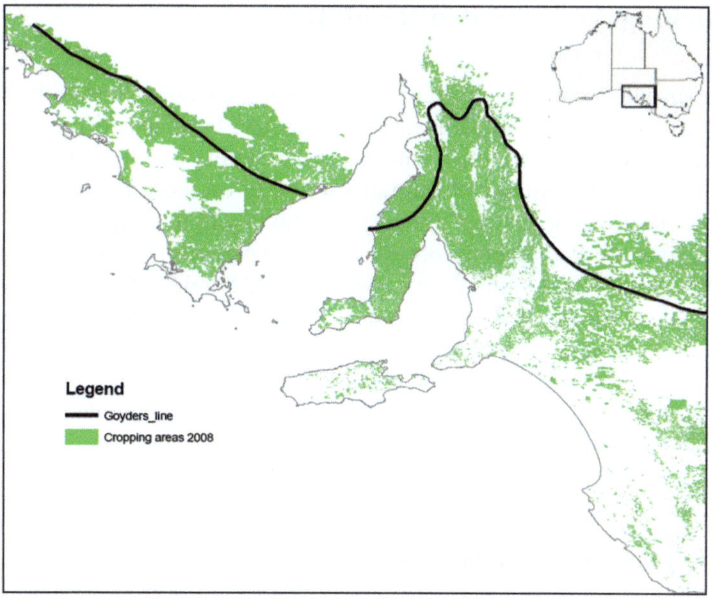

Fig. 2 Goyder's line of *reliable rainfall* [5]

2.1 Goyder's Line of Reliable Rainfall

In South Australia, a line drawn by the Surveyor General George Goyder in 1865, believed to indicate the edge of the area suitable for agriculture, is still regarded as important in the discussion of land-use planning in South Australia over 150 years later. Figure 2 shows a map of Goyder's line including cropping areas (for 2008 which is relatively consistent year to year). This is known as the South Australian agricultural cropping belt. From the figure, we can see that cropping has extended north of the Goyder's line and in other areas cropping south of the line (only during exceptionally good seasons) [4, 5, 7, 10].

Recent research on climate change projections for the dry-land agricultural zones of South Australia suggests that annual average temperatures may increase by 0.4 to 1.8 °C and we may see decreases by as much as 15% in average annual rainfall by 2030 [14]. The projected declines, especially in spring, will have a significant impact on current agriculture practices. Researchers are interested in investigating the potential impact of projected climate change on the position of Goyder's line, thereby potentially reducing the areas where reliable agriculture productivity can be maintained in South Australia [7]. Our main aim is to consider how simulated rainfall generated by models using multidimensional copulas can assist water management in the South Australian agricultural cropping belt and give farmers greater understanding of the likely impact of variations in spring rainfall.

3 Modelling Accumulated Rainfall

For future management of agriculture and assessing risk associated with a changing climate, we require rainfall totals at various timescales as input to ecological models. Generated synthetic rainfall totals with realistic monthly and seasonal statistics are essential to assess the impact of rainfall variability on these systems and for risk assessment. To drive these simulations, it is necessary to develop stochastic rainfall simulations that accurately reflect observed rainfall patterns [11]. To demonstrate the proposed models, we use rainfall totals between 1882 and 2006 from the Australian Bureau of Meteorology for seasonal rainfall in spring for rainfall stations around Goyder's line—at stations 019017 Hawker above the line, 019005 Orroroo (Black Rock) on Goyder's line and 021010 Clare (Brinkworth–Bungaree) below the line. Figure 3 shows that Clare has the highest annual rainfall, which decreases steadily as we move north toward Hawker [7].

3.1 The Gamma Distribution

The Gamma distribution has been widely used to model both short-term and long-term rainfall accumulations at a specific location. For each fixed timescale, there

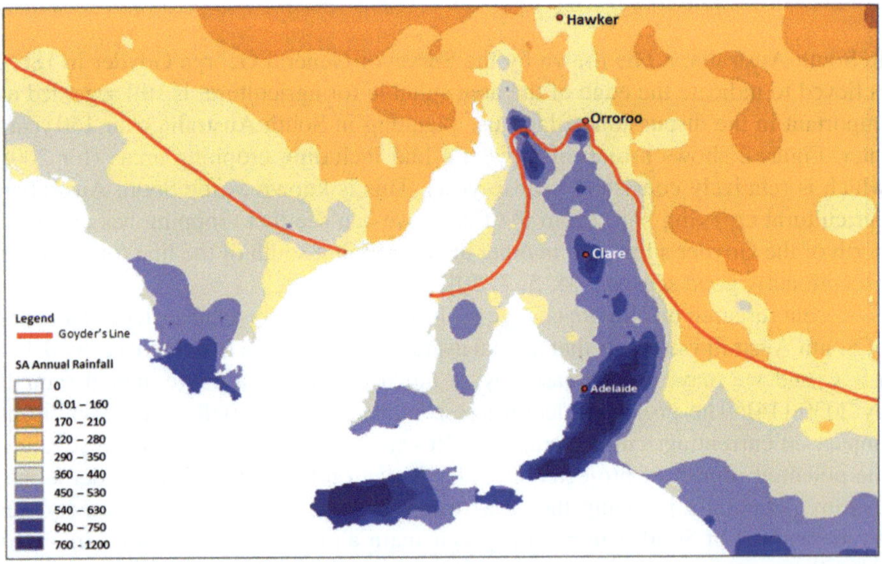

Fig. 3 South Australia rainfall map [7]

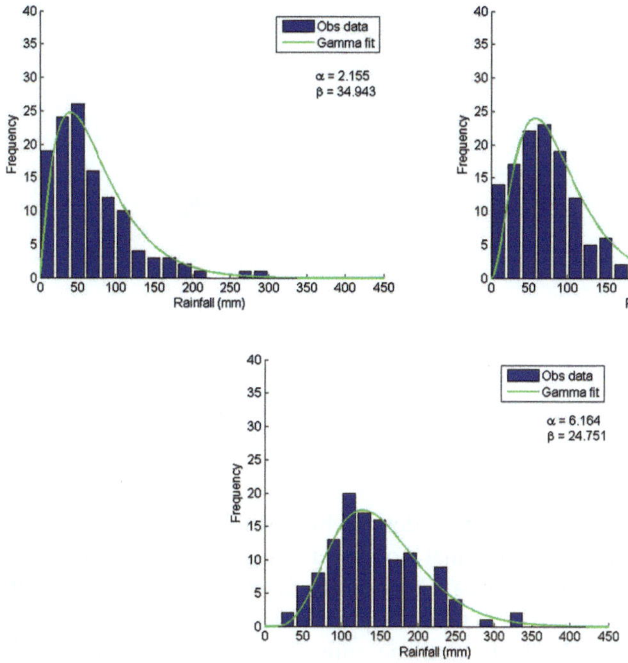

Fig. 4 Observed spring rainfall and Gamma fit for Hawker (top left), Orroroo (top right) and Clare (bottom)

is general agreement that a mixed gamma distribution with cumulative distribution $F_{p_0,\alpha,\beta}(x) = P[0 \le X < x]$ for $x \ge 0$ given by

$$F_{p_0,\alpha,\beta}(x) = p_0 + (1 - p_0) \int_0^x f_{\alpha,\beta}(x)dx$$

where $\alpha > 0$ and $\beta > 0$ are parameters determined by maximum likelihood estimation and the probability density function is defined for $x > 0$ by the formula

$$f_{\alpha,\beta}(x) = \frac{x^{\alpha-1}e^{-x/\beta}}{\beta^\alpha \Gamma(\alpha)}.$$

Figure 4 shows histograms for observed spring rainfall at Hawker (top left), Orroroo (top right) and Clare (bottom) with fitted Gamma distributions. A Kolmogorov–Smirnov test shows that the Gamma distributions provide a good fit to the observed seasonal total.

There is no natural joint distribution with correlated marginal gamma distributions for rainfall on each separate month and an appropriate gamma distribution for the total season rainfall. Consequently, more sophisticated models have been proposed

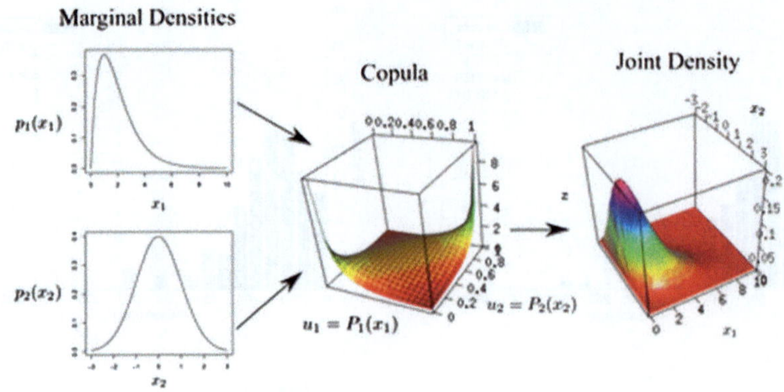

Fig. 5 A joint probability distribution with prescribed marginal densities [6]

[2, 11–13]. Copulas of maximum entropy defined by multi-stochastic matrices were first proposed in [11] to preserve marginal monthly distributions and match observed non-zero correlations. In simple terms, *Copulas* are functions that join (or couple) multivariate distribution functions to their marginal distributions (Fig. 5).

The essence of the copulas is that one measures marginal distributions of multivariate distributions, which must be determined in order to make predictions. To model the joint probability distribution for a vector-valued random variable $\boldsymbol{X} = (X_1, X_2, \ldots, X_m) \in (0, \infty)^m$ with known marginals $u_i = F_i(x_i)$ we simply construct uniformly distributed random variables $U_i = F_i(X_i) \in (0, 1)$ for each $i = 1, 2, \ldots, m$ and use the m-dimensional copula

$$C(\boldsymbol{u}) = C(\boldsymbol{F}(\boldsymbol{x})) = C(F_1(x_1), F_2(x_2), \ldots, F_m(x_m)).$$

By an elementary transformation on each marginal distribution, the copula can be used to find a corresponding joint probability distribution for seasonal rainfall with prescribed marginal Gamma distributions ($F_i(x_i)$) for each of the monthly rainfalls and specified correlation coefficients. The techniques include representation of multivariate distributions for correlated monthly and seasonal rainfall using multidimensional copulas with maximum entropy. A comprehensive discussion of the choice of appropriate copulas for rainfall modelling and simulation can be found in [1, 2, 11–13].

3.2 Results

We observe that generated monthly rainfall totals modelled as mutually independent gamma random variates when accumulated generate seasonal totals with significantly

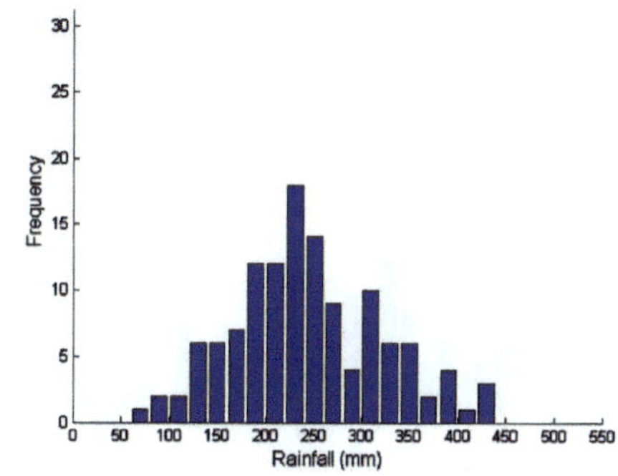

a) Time:125years(255,5939)

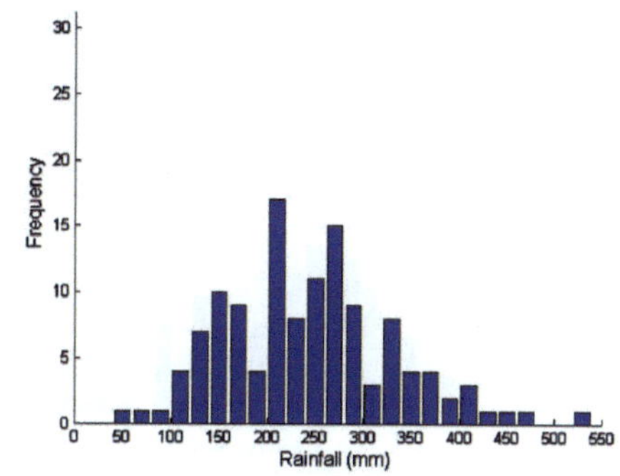

b) Time:125 years (254,7710)

Independence

Maximum Entropy

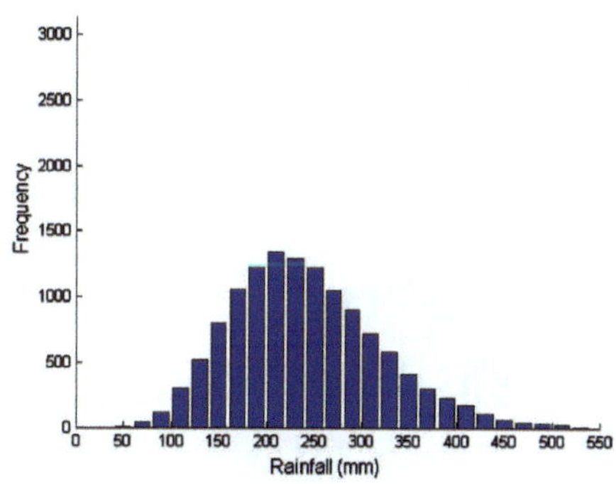

c) Time:12500years(252,6139)

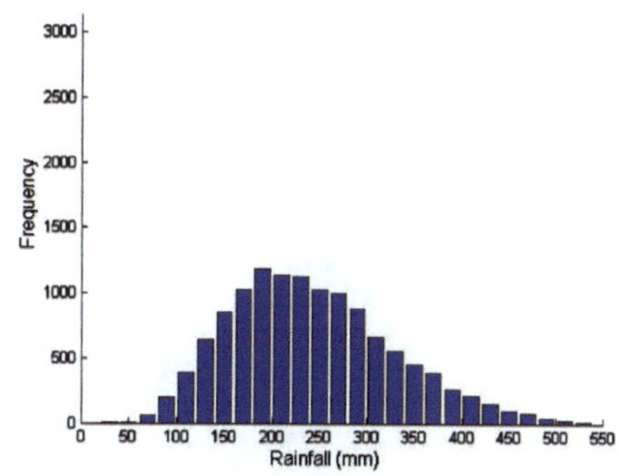

d) Time:12500 years (253,7534)

Fig. 6 Simulations: growing season rainfall at Orroroo. Observed statistics: $\mu = 253$, $\sigma^2 = 7675$ (μ, σ^2)

lower variances than observed in the historical records. This is clearly illustrated in Fig. 6a, c for seasonal rainfall at Orroroo.

Similarly for Hawker and Clare, we note that, according to simulations in which the monthly rainfall totals are modelled as independent gamma distributions, the accumulated seasonal variance is too low. Refer to Fig. 7 a, c for season rainfall at Hawker and Fig. 8a, c for season rainfall at Clare. The variance of the synthetic totals increases if the model incorporates correlation between individual months totals. Refer to Fig. 6b, d for season rainfall at Orroroo, Fig. 7b, d for season rainfall at Hawker and Fig. 8b, d for season rainfall at Clare.

For the large samples of seasonal totals generated by the multidimensional copula of maximum entropy, the standard deviation is quite close to the observed standard deviation. It is important to note that although we have chosen to use marginal Gamma distributions in this study, the multidimensional copula of maximum entropy can be

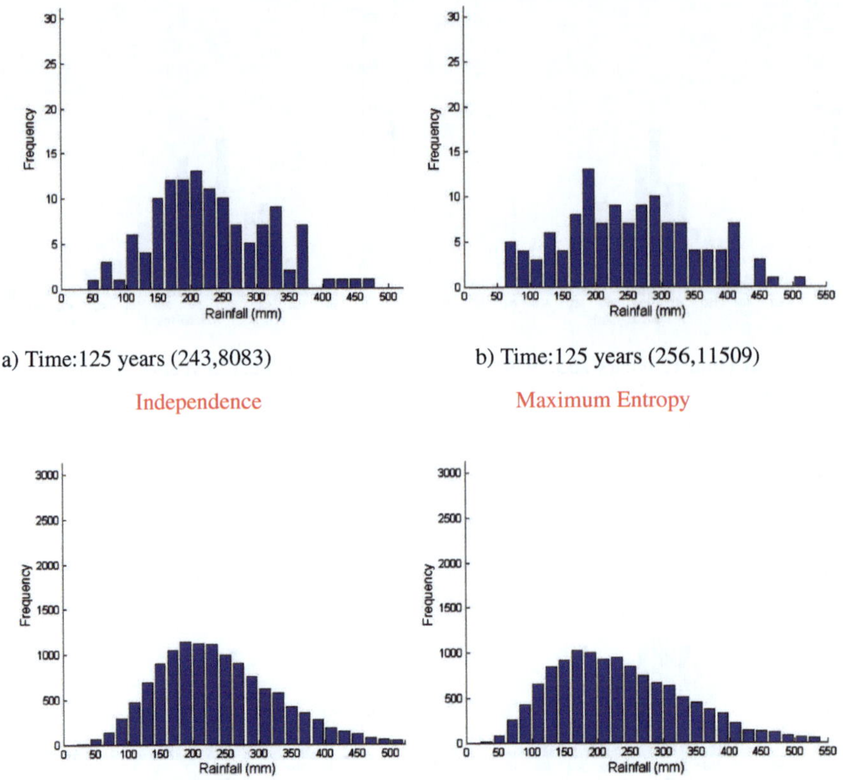

a) Time:125 years (243,8083)

b) Time:125 years (256,11509)

c) Time:12500 years (248,8599)

d) Time:12500 years (246,10802)

Fig. 7 Simulations: growing season rainfall at Hawker. Observed statistics: $\mu = 247, \sigma^2 = 10620$ (μ, σ^2)

used with any appropriate marginal distributions. To highlight the inherent variability in samples of this size, we used the multidimensional copula to simulate seasonal rainfall over a period of N = 125 years (the same number of years of the observed records). We note that each sample is generated by the same underlying distribution and yet the sample statistics show a high degree of variation. Repeated simulations confirm that the Kolmogorov–Smirnov test statistics for the distributions generated by the multidimensional copula of maximum entropy and the observed distribution lie within the acceptable limits defined by the 5% significance level.

Our main aim is to consider how simulated rainfalls generated by models using multidimensional copulas might change if the key parameters change. By changing these parameters, we can assess the potential impact of projected climate change and the negative implications for agriculture in South Australia. For the purpose of this

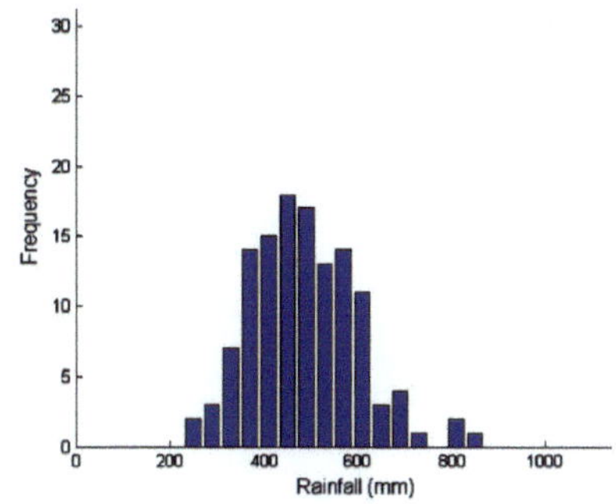

a) Time:125 years (505,13129)

Independence

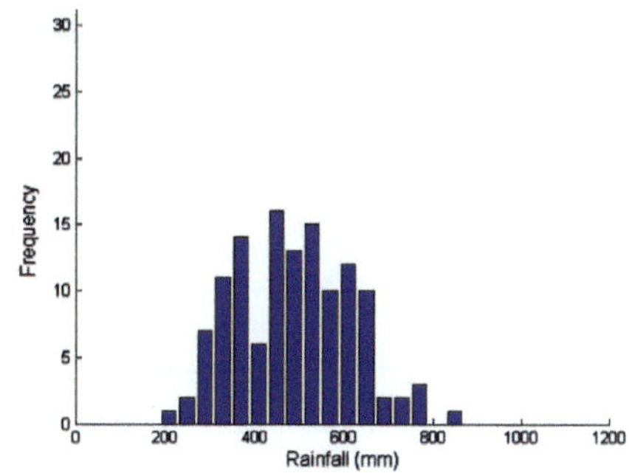

b) Time:125 years (508,16257)

Maximum Entropy

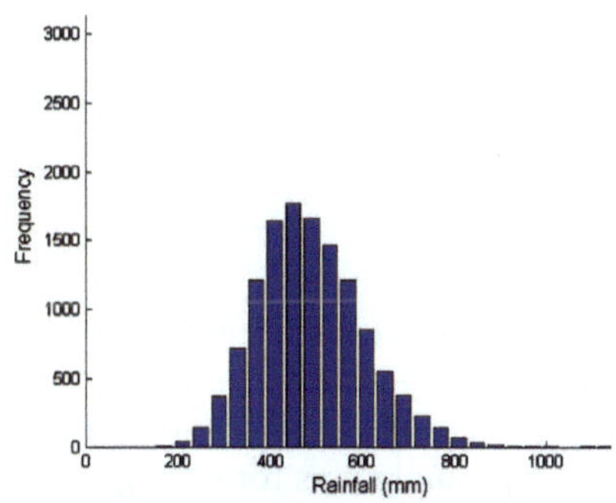

c) Time:12500 years (508,13203)

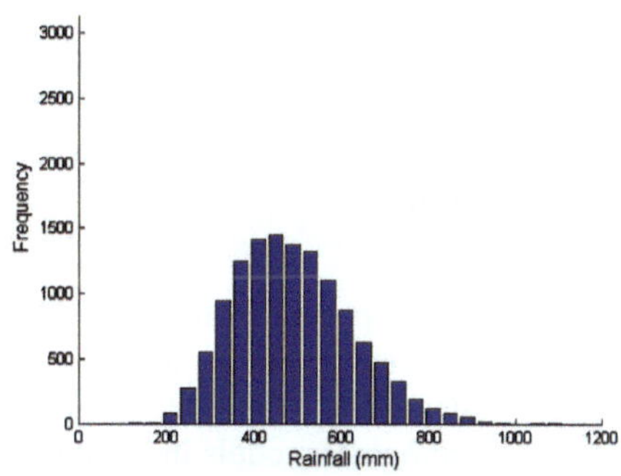

d) Time:12500 years (508,17893)

Fig. 8 Simulations: growing season rainfall at Clare. Observed statistics: $\mu = 508$, $\sigma^2 = 18160$ (μ, σ^2)

investigation, we focus on wheat grain yields as wheat is traditionally the primary crop in South Australia's agricultural belt. In Fig. 9, we compare the average grain yield in tonnes per hectare (t/ha) for various climate scenarios, namely, historical climate, 5, 10 and 20% reduction in rainfall [8]. For the purpose of this study, the climate change scenarios we focus on decreases in rainfall. Future studies could also include elevated temperature and CO_2 levels.

Within an already dry South Australian climate, a further reduction in seasonal totals will certainly impact on current whcat production. Marginal areas such as Orroroo (located on Goyder's Line) will be significantly impacted and current practices can no longer be maintained.

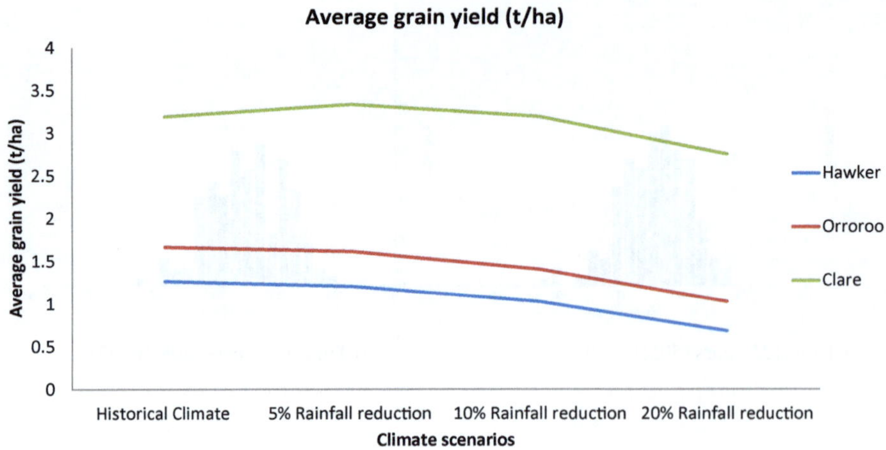

Fig. 9 The average grain yield (tonne per hectare) for various climate scenarios

4 Conclusions

In this study, we present a model using a multidimensional copula of maximum entropy to generate synthetic seasonal totals for three locations in South Australia around Goyder's line of reliable rainfall. The simulated rainfall totals can be used as input to ecological models including yield crop simulations for management and risk assessment by changing key model parameters. Farmers and land managers need to consider alternative practices and reliable agriculture land management in an uncertain and highly variable climate.

References

1. J. Boland, P. Howlett, J. Piantadosi, R. Zakaria, Modelling and simulation of volumetric rainfall for a catchment in the Murray-Darling Basin. ANZIAM J **58**(2), 119–142 (2016)
2. J. Borwein, P. Howlett, J. Piantadosi, Modelling and simulation of seasonal rainfall using the principle of maximum entropy. Entropy **16**, 747–769 (2014)
3. Bureau of Meteorology, Australian Government Bureau of Meteorology (2016)
4. R.J. French, Changes in cropping systems at the boundaries of the pastoral and cropping zones in Southern Australia. Rangel. J **15**, 117–132 (1993)
5. P. Hayman, B. Alexander, U. Nidumolu, N. Wilhelm, Using spatial and temporal analogues to consider impacts and adaptation to climate change in the South Australian grain belt, in *Proceedings of 15th Australian Agronomy Conference*, Lincoln, New Zealand (2010)
6. J.M. Hernandez-Lobato, A. Suarez, Semiparametric bivariate Archimedean copulas. Comput. Stat. Data Anal. **55**(6), 2038–2058 (2011)
7. S.M. Howden, P. Hayman, The distribution of cropping under climate change: Goyders line, in *Proceedings of the Greenhouse Conference*, Melbourne, Australia, 14–17 Dec 2005

8. C. Liddicoat, P. Hayman, B. Alexander, J. Rowland, D. Maschmedt, M-A. Young, J. Hall, T. Herrmann, S. Sweeney, Climate change, wheat production and erosion risk in South Australias cropping zone: linking crop simulation modelling to soil landscape mapping, Government of South Australia, through Department of Environment, Water and Natural Resources (2012)
9. R.B. Nelsen, *An Introduction to Copulas*, (Springer, New York, 1999), pp. 139
10. U. Nidumolu, P. Hayman, S. Howden, B. Alexander, Re-evaluating the margin of the South Australian grain belt in a changing climate. Clim. Res. **51**, 249–260 (2012)
11. J. Piantadosi, J. Boland, P. Howlett, Generating synthetic rainfall on various time scales—daily, monthly and yearly. Environ. Model. Assess. **14**(4), 431–438 (2009)
12. J. Piantadosi, P. Howlett, J. Borwein, Copulas of maximum entropy. Optim. Lett. **6**(1), 99–125 (2012)
13. J. Piantadosi, R.S. Anderssen, Rainfall simulation from an inverse problems perspective, in *MODSIM2015, 21st International Congress on Modelling and Simulation*, ed. by T. Weber, M.J. McPhee, R.S. Anderssen (Modelling and Simulation Society of Australia and New Zealand, 2015), pp. 133–137. ISBN: 978-0-9872143-5-5
14. PIRSA, *Adapting to Climate Change* (Primary Industries and Resources SA, 2011)

Vertex Representation of Convex Hulls for Fast Evaluation of Boundary Constraints in Model-Based Calibration for Automotive Engines

Florin Nae and Hayato Waki

Abstract In the field of automotive engines calibration, generating a boundary model of an admissible operation domain, for instance, a convex hull, is a step often required when solving constrained optimization problems addressed by Model-Based Calibration (MBC) software suites, such as Model-Based Calibration Toolbox from MathWorks. However, depending on the amount of data modeled, generating a convex hull might become a computationally intensive process. This is due to the fact that the half-space representation of the convex hull is used. We discuss here an alternative representation of the convex hull, the vertex representation, which proves capable to reduce the computational cost for specific conditions. Numerical comparisons in this article are executed in MATLAB using MBC Toolbox commands and show that for a representative optimization problem, the vertex representation outperforms the half-space representation.

Keywords Convex hull · Boundary modeling · Optimization
Model-based calibration · Automotive engine · MBC toolbox

F. Nae (✉)
The MathWorks G.K. Nagoya Office, 16/F KDX Sakuradori Bldg.,
3-20-17, Marunouchi, Naka-ku, Nagoya, Aichi 460-0002, Japan
e-mail: Florin.Nae@mathworks.co.jp

H. Waki
Institute of Mathematics for Industry, Kyushu University,
744 Motooka, Fukuoka, Nishi-ku 819-0395, Japan
e-mail: waki@imi.kyushu-u.ac.jp

R. S. Anderssen et al. (eds.), *Agriculture as a Metaphor for Creativity in All Human Endeavors*, Mathematics for Industry 28,
https://doi.org/10.1007/978-981-10-7811-8_10

1 Introduction

Model-Based Calibration (abbr. MBC) is a systematic approach that allows cost-effective and accelerated development of automotive engines. At the same time, MBC enables engineers to design fuel-efficient and eco-friendly engines.

To allow the engine to perform at optimized operating conditions, regression-based response models of various engine characteristics—such as torque, fuel consumption, or exhaust emissions—are created in MBC as functions of controllable engine inputs such as airflow or fuel flow, and, next, optimization techniques are applied to the fitted models to obtain best values for the controllable inputs [5]. In this way, optimal parameters needed by the control algorithms used in the electronic engine control unit can be efficiently calculated using a model-based design approach.

Boundary modeling is used in MBC to represent or approximate a region where the automotive engine works normally. We call this region the admissible operation domain (abbr. AOD). In general, as it is assumed that internal combustion engines are highly nonlinear systems, it is impossible to exactly represent the AOD from a finite number of acquired data. Instead, one approximates the AOD. One of the approximations of AOD is the convex hull of a finite set of points, implemented in typical MBC software such as [5]. One reason for choosing the convex hull is that optimization routines usually deal better with convex regions than with nonconvex ones. After approximating the AOD, this is used as a constraint in constrained optimization problems, to estimate better control parameters for automotive engines. This is why a proper handling of AODs is very important in engine optimization problems.

The contribution of this article is to introduce the approach presented in [6] and apply it to a different problem than the one solved in [6], to further validate the method. In particular, we discuss the performance of the convex hull approximation of the AOD in the response surface methodology for a diesel engine data set. The motivation stems from the limitations of the convex hull methodology implemented in MBC software, such as [5]. These can become severe for particular problems—for instance, diesel engine data consisting of a large number of variables and measurements—since a convex hull cannot even be generated due to large memory requirements or because it takes an unreasonably long time to generate it.

It must be noted that two representations for the convex hull of a set of points are possible, the half-space representation (abbr. H-repr) and the vertex representation (abbr. V-repr). These representations are discussed in [2, 4]. The main reason for the numerical difficulties encountered when generating convex hulls was discussed in [6] and is related to the fact that the H-repr for the convex hull of a set of points is typically used, instead of the V-repr. The findings reported in [6] show that, for specific conditions, the V-repr outperforms the H-repr in the numerical experiments performed. In the present article, we validate the approach proposed in [6] by considering a different objective for the optimization problem and performing additional numerical studies.

In the next section, we provide the numerical comparison of the V-repr of the convex hull of the AOD with the H-repr. Throughout this article, we assume that the measured engine data was acquired by keeping the engine under test at steady condition by controlling its inputs on an engine bench. We provided a brief introduction of the boundary modeling in model-based calibration in [6]. In addition, see [2, 4, 6] for the detail of the convex hull of a set of points.

2 Application of the Convex Hull: An Optimization Problem in the MBC Response Surface Methodology

2.1 *Convex Hull and Optimization Problems in MBC*

For a finite set V of data $\mathbf{v}_i \in \mathbb{R}^n$ $(i = 1, \ldots, m)$ obtained from some of the processes in MBC, the convex hull can be formulated as follows:

$$\text{conv}(V) := \left\{ \mathbf{v} \in \mathbb{R}^n : \mathbf{v} = \sum_{i=1}^{m} \alpha_i \mathbf{v}_i, \sum_{i=1}^{m} \alpha_i = 1, \alpha_i \geq 0 \ (i = 1, \ldots, m) \right\}.$$

This is called the vertex representation (hereafter abbr. V-repr) of V. Minkowski-Weyl's theorem ensures that there exist $\mathbf{a}_i \in \mathbb{R}^n$ and $b_i \in \mathbb{R}$ for $i = 1, \ldots, p$ such that

$$\text{conv}(V) = \left\{ \mathbf{v} \in \mathbb{R}^n : \mathbf{a}_i^T \mathbf{v} \leq b_i \ (i = 1, \ldots, p) \right\}.$$

This is called the half-space representation (hereafter abbr. H-repr) of V.

The convex hull approximation $\text{conv}(V)$ of the AOD is used for optimization in MBC, using for example the response surface methodology. In particular, we need to solve an optimization problem of an objective function over the approximated AOD or a subset of the AOD, thus the AOD acting as a constraint for the optimization problem. This is mathematically formulated as

$$\min_{\mathbf{v} \in \mathbb{R}^n} \left\{ f(\mathbf{v}) : g_j(\mathbf{v}) \geq 0 \ (j = 1, \ldots, k), \mathbf{v} \in \text{conv}(V) \right\}, \tag{1}$$

where $f(\mathbf{v})$ is the objective function and $g_j(\mathbf{v}) \geq 0$ is an engine operating constraint. We have the two reformulations of (1). The first one is by the H-repr of the convex hull $\text{conv}(V)$, and (1) is formulated by

$$\min_{\mathbf{v} \in \mathbb{R}^n} \left\{ f(\mathbf{v}) : g_j(\mathbf{v}) \geq 0 \ (j = 1, \ldots, k), \mathbf{a}_i^T \mathbf{v} \leq b_i \ (i = 1, \ldots, q) \right\}, \tag{2}$$

where is q is the number of the generated half-spaces. The second one by the V-repr of the convex hull $\text{conv}(V)$ has the V-repr, and (1) is formulated by

$$\min_{\alpha_1,\ldots,\alpha_m \in \mathbb{R}} \left\{ \tilde{f}(\alpha_1,\ldots,\alpha_m) : \begin{array}{c} \tilde{g}_j(\alpha_1,\ldots,\alpha_m) \geq 0 \ (j = 1,\ldots,k), \\ \sum_{i=1}^{m} \alpha_i = 1, \alpha_i \geq 0 \ (i = 1,\ldots,m) \end{array} \right\}, \tag{3}$$

where $\tilde{f}(\alpha_1,\ldots,\alpha_m) = f\left(\sum_{i=1}^{m} \alpha_i \mathbf{v}_i\right)$ and $\tilde{g}_j$ is defined in a similar manner to $\tilde{f}$.

Before discussing a numerical comparison of V-repr with H-repr, we mention some advantages and disadvantages of the V-repr formulation:

(I) We can skip the process of construction of $\mathbf{a}_i$ and b_i of conv(V). As we have already seen in [6, Table 1], this is computationally intensive, and thus one can greatly reduce the computational cost.

(II) The number of inequality constraints in V-repr is much less than H-repr. In general, the H-repr has exponentially increasing number of many linear inequalities of type $\mathbf{a}_i^T \mathbf{v} \leq b_i$. An example is provided below in Sect. 2.2.

(III) On the other hand, the number of the variables in V-repr increases. In fact, it is m (the number of points in V), while for H-repr is n (number of dimensions). This is the disadvantage of the V-repr formulation.

2.2 Numerical Experiment for a Diesel Engine Data Set

To compare the H and V representations, we use a diesel engine data set being collected using a space-filling Design of Experiment (DOE) based on a Sobol pseudorandom sequence. This data set is denoted by V and consists of 875 observations, where at each measured observation the following engine signals were measured:

MAINSOI : start of main fuel injection event [degCA],
FUELPRESS : common rail fuel pressure [MPa],
VGT : variable-geometry turbo charger,
VGTPOS : vane position [mm],
EGR : exhaust gas recirculation,
EGRPOS : valve opening position [ratio],
MAINFUEL : amount of injected fuel mass during main injection event [mg/stroke],
EGRMF : mass flow ratio of recirculated exhaust gas [ratio],
AFR : air–fuel ratio [ratio],
VGTSPEED : VGT rotational speed [rpm],
PEAKPRESS : in-cylinder peak pressure [MPa], and
BSNOX : brake-specific NO_x [g/kWh].

Moreover, the measurements were performed by keeping the engine under test at seven specific engine operating points, expressed as (Engine Speed SPEED [rpm], Brake Torque BTQ [Nm]) pairs. In this way, for each operating point defined by a

SPEED—BTQ pair, the rest of the signals were measured. Around 120 measurements were performed for each operating point. Next, for each of these engine operating points, response surface models were generated using as model inputs some subsets of the measured signals listed above. In such a way, seven different models were generated, one for each operating point. The necessity to generate multiple models comes from the fact that modern diesel engines are highly nonlinear complex systems and fitting a single response model for all operating conditions has become increasingly difficult in recent years, leading to the necessity to break down the problem in order to solve it.

The response model considered as the objective for the optimization problem in this investigation is BSNOX, a measure of NO_x, which is one of the most dangerous engine's exhaust emissions. NO_x is a generic term for the nitrogen oxides that are most relevant for air pollution, namely nitric oxide (NO) and nitrogen dioxide (NO_2). We have chosen BSNOX due to the high interest in engine exhaust emissions estimation and control. This interest is related to the increasingly severe environmental regulations all over the world.

To investigate the V-repr and H-repr of convex hulls, three sets of models were generated, by using the following three sets of inputs:

(A type) BSNOX_p(MAINSOI, FUELPRESS, VGTPOS, EGRPOS),
(B type) BSNOX_p(MAINSOI, FUELPRESS, VGTPOS, EGRPOS, MAINFUEL, EGRMF, AFR), and
(C type) BSNOX_p(MAINSOI, FUELPRESS, VGTPOS, EGRPOS, MAINFUEL, EGRMF, AFR, VGTSPEED, PEAKPRESS),

where $p = 1, \ldots, 7$ represents the seven engine operating conditions considered in this study.

The dimension n of these data sets used in the three models is 4, 7, and 9, respectively. We considered different n in order to investigate the scalability of V-repr and to compare the computational cost with H-repr. It should be noted that for modern diesel engines a number of 10 or even more controllable inputs are becoming common and this is the main cause of the previously mentioned numerical problems related to the generation of convex hulls.

For each data set, we have solved the following seven optimization problems, one for each operating point set:

$$\min_{\mathbf{v}\in\mathbb{R}^n} \left\{ f_p(\mathbf{v}) : \mathbf{v} \in \text{conv}(V) \right\} \ (p = 1, \ldots, 7),$$

where V consists of a subset of the initial 875 n-dimensional vectors, since the approach we adopted is a point-by-point one. The measured points in each subset are unique. As an indication, each local model consisted of 125 of such measurements, and for each local model a corresponding convex hull was generated.

Various candidate response models were investigated in this study:

1. Second- and third-order polynomials,
2. Quadratic RBF models, that consisted of a Radial Basis Function (RBF) network having a multiquadric kernel, plus a quadratic term, and
3. Gaussian process models, that used an automatic relevance determination (ARD) squared exponential kernel and a constant basis.

For each local BSNOX_p $(p = 1, \ldots, 7)$ model, above-listed candidate models were generated and the one having the smallest leave-one-out root-mean-squared error was selected as the best candidate. For this particular problem, the quadratic RBF and Gaussian process models proved to be the best ones. The polynomial models were unable to capture the highly nonlinear behavior of the BSNOX response.

Next, an optimization problem was considered. For this problem, we generate seven objective functions f_p (using BSNOX_p as the objective to be minimized) and we add $\mathbf{l} \leq \mathbf{v} \leq \mathbf{u}$ as constraints $g_j(v) \geq 0$. Here, $\mathbf{l}$ and $\mathbf{u}$ are obtained from the measurement data. In particular, the objective functions f_p are smooth but nonconvex, and thus the problem becomes a nonlinear nonconvex optimization problem.

A point-by-point minimization problem for BSNOX was performed and Table 1 displays a numerical comparison between the two methods. In this numerical experiment,[1] we use MathWorks Model-based Calibration Toolbox [5] to obtain the BSNOX models, and MathWorks Optimization Toolbox and Global Optimization Toolbox products for the optimization task. We compare computation time for the two methods. The third column in Table 1 is the total computation time for the optimization. The forth column in Table 1 stands for the number of constraints in the optimization. Note that these numbers for (2) are the number of linear constraints defining the convex hull's polyhedron in $\mathbb{R}^n$. For V-repr, no convex hull is constructed so we use "–" in Table 1.

Table 1 Computation time[s] comparison of V-repr with H-repr

Construction of H-repr			Optimization	# of constraints
(2)	(A type)	0.12	49.59	1,834 linear ineq.
	(B type)	10.35	60.84	211,299 linear ineq.
	(C type)	363.52	547.48	3,198,561 linear ineq.
(3)	(A type)	–	434.46	8 linear ineq. and 1 linear eq.
	(B type)	–	420.65	14 linear ineq. and 1 linear eq.
	(C type)	–	466.08	19 linear ineq. and 1 linear eq.

[1] The specification on the used computer is as follows: OS is Windows 7, the CPU is Intel®Core™i7-5600U with 2.60 GHz, and the memory is 12 GB and version of MATLAB is R2016b.

To solve the optimization problem, we considered a two-step approach. In the first step, we used the pattern search method [1], which is a derivative-free method, to find an initial solution, and next we used this solution as the starting point for the interior-point method, in order to refine the solution. This strategy proved to be efficient since, even if pattern search was found to give solutions that satisfied the constraints tolerance, applying interior-point method was found to further increase the accuracy of the solution. By using this strategy, similar results were obtained for both H-repr and V-repr (to be more specific, numerical difference between the two solutions being less than 10^{-3} [g/kWh]), thus validating the V-repr method with respect to the "traditional" H-repr. The optimization settings that were used to obtain the solution are listed in Tables 2 and 3. For the settings not listed in Tables 2 and 3, the default settings were used. We observe the following from Table 1.

(i) Since the complexity of optimization methods depends on the length of the input size (which is related to the number of variables and nonzero coefficients in the constraints), in (A type) and (B type), i.e., for $n = 4$ and 7, H-repr is faster than V-repr, whereas for (C type) optimization of V-repr is already faster than H-repr, even if we do not consider the additional time needed to create the H-repr. In fact, as it can be seen from Table 1, the number of linear inequalities in H-repr increases considerably with the number of dimensions. Consequently,

Table 2 Representative options used by `patternsearch`. Default values were used for the rest of the options not listed here

Pattern	GPS Positive basis 2N
Maximum number of function evaluations	$200\times$ (the number of variables)
Mesh tolerance	10^{-8}
Step tolerance for free variables	10^{-6}
Constraint violation tolerance	10^{-8}
Objective function tolerance	10^{-8}

Table 3 Optimization options used by the interior-point method. Default values were used for the rest of the options not listed here

Maximum number of function evaluations	5000 for (2). 20000 for (3)
Maximum number of iterations	500
Maximum change in variables for finite-difference gradients	0.1
Minimum change in variables for finite-difference gradients	10^{-8}
Step tolerance for free variables	10^{-6}
Constraint violation tolerance	10^{-8}
Objective function tolerance	10^{-6}

the evaluation of computed solutions at each iteration becomes computationally intensive. In contrast, the number of variables in (2) is n, while it is m in (3). Thus, solving optimization problems by H-repr is much faster than by V-repr for (A type) and (B type).

(ii) The computation time of constructing conv(V) into the form of the H-repr considerably increases as n increases. This is related to the number of faces of the polyhedron used in the H-repr, which increases exponentially with the number of dimensions, as can be seen in Table 1. As (3) can skip this conversion, we can expect that, for the present problem, (3) is more efficient than (2) for $n \geq 9$.

3 Conclusion

In order to further validate the approach proposed in [6], we have applied the same methodology to a different problem. Similar to [6], for the examples investigated we observe that, when the dimension n of space in which a set V of points lies is more than seven, the computational cost can be reduced when using the V-repr of a set of points instead of the H-repr. An additional benefit is that in the V-repr there is no need to construct a convex hull. This benefit becomes crucial for data consisting of a large number of variables and measurements—in such cases, a convex hull cannot even be generated due to large memory requirements or because it can take an unreasonably long time to generate it—so the V-repr becomes maybe the only method that can be used to define constraints related to AOD. The problem addressed in the present contribution is more challenging numerically, due to the non-smooth behavior of the objective function used in the optimization problem. To solve it, we had to use a two-step optimization strategy, where the solution obtained using a derivative-free method (pattern search, a particular form of direct search) was further refined using an interior-point method. In spite of the increased numerical difficulty, results validate the proposed approach, confirming that V-repr can be used to define an AOD faster than the traditional H-repr, for problems with an increased number of dimensions. This enables solving in an accurate yet fast way the increasingly complex optimization problems related to modern internal combustion engines.

Acknowledgements This article is based on discussion in workshops of the IMI Joint Research Projects "Research of Boundary Modeling" (Principal investigator: Mr. Satoru Watanabe, Toyota Motor Company) and "Research of Engine Calibration from the Viewpoint of Mathematical Science" (Principal investigator: Mr. Masahiro Satou, Honda Automobile R&D).

References

1. C. Audet, J.E. Dennis Jr., Analysis of generalized pattern searches. SIAM J. Optim. **13**(3), 889–903 (2003)
2. A. Barvinok, *A Course in Convexity*. Graduate Studies in Mathematics, vol. 54 (American Mathematical Society, 2002)
3. V. Chvátal, *Linear Programming* (W.H. Freeman & Co. Ltd., 1983)
4. B. Grünbaum, *Convex Polytopes* (Springer, New York, 2013)
5. Model-Based Calibration Toolbox, MathWorks (2016), http://www.mathworks.com/products/mbc. Accessed 25 Apr 2016
6. H. Waki, F. Nae, Boundary modeling in model-based calibration for automotive engines via the vertex representation of the convex hulls **9**(1), 1 (2017)

Mathematical Modeling of Rock Pore Geometry and Mineralization: Applications of Persistent Homology and Random Walk

Takeshi Tsuji, Fei Jiang, Anna Suzuki and Tomoyuki Shirai

Abstract Mathematical methods used to model heterogeneous pore geometry of natural rocks and their temporal evolution (mineralization processes) are explored. Recent development of X-ray microcomputed tomography enables high-resolution (micrometers) pore geometry of rock to be obtained. Nevertheless, exploring the complex spatial distribution of pore bodies, and relating this information to hydraulic and elastic properties, remains a challenge. In this study, persistent homology is first applied to describe heterogeneous rock pores, which captures the appearance and disappearance of topological features. The persistence diagram derived from this analysis shows the characteristic features of rock pore. Next, random walk is used to model rock mineralization processes. The results show that rock pore evolution is successfully modeled using random walk by defining the probability of mineral precipitation and dispersion degree in each grid cell of a modeled rock body. The mineralization parameter can be flexibly changed and a short computation time used when using random walk; this approach may thus be practical when simulating rock evolution processes such as long-term chemical reactions in a reservoir.

T. Tsuji (✉) · F. Jiang
International Institute for Carbon-Neutral Energy Research, Kyushu University, Fukuoka, Japan
e-mail: tsuji@mine.kyushu-u.ac.jp

T. Tsuji
Department of Earth Resources Engineering, Kyushu University, Fukuoka, Japan

F. Jiang
Yamaguchi University, Yamaguchi, Japan
e-mail: fjiang@yamaguchi-u.ac.jp

A. Suzuki
Institute of Fluid Science, Tohoku University, Sendai, Japan
e-mail: anna.suzuki@tohoku.ac.jp

T. Shirai
Institute of Mathematics for Industry, Kyushu University, Fukuoka, Japan
e-mail: shirai@imi.kyushu-u.ac.jp

R. S. Anderssen et al. (eds.), *Agriculture as a Metaphor for Creativity in All Human Endeavors*, Mathematics for Industry 28,
https://doi.org/10.1007/978-981-10-7811-8_11

Keywords Natural rock · Heterogeneous pore geometry · Persistent homology
Rock mineralization · Random walk

1 Introduction

In oil or gas production, CO2 geological storage, and geothermal power, it is important to quantify the hydraulic and elastic properties of the natural rocks being exploited. The porosity, defined as the fraction of pore volume per total volume, of a typical reservoir is about 20%, suggesting that there is significant space in the subsurface geology. Fluid flow within the pore spaces has a strong influence on the production of hydrocarbons (oil and gas) or efficiency of CO2 injection into any given reservoir. Permeability (K) is the ability of a porous material to allow fluids to pass through it and can be used to predict fluid migration in a reservoir [5]. Permeability can be defined as follows:

$$K = \mu \frac{\nu}{\Delta P/\Delta x}, \tag{1}$$

where μ is the viscosity of the fluid, ν is the fluid flow velocity, and $\Delta P/\Delta x$ is the pressure gradient within the medium. The elastic properties of the rock, such as elastic moduli, are used in calculation of seismic velocities, such as P- and S-wave velocities. The elastic properties are key parameters in geophysical exploration of geological structures, such as oil reservoirs [22].

Both permeability and elastic properties are governed by volumetric and geometric considerations, whereas porosity is a volumetric description. Thus, both hydraulic and elastic properties are dependent on the geometry of microscopic-scale pore rock (Fig. 1). For example, hydraulic properties such as permeability have been described by the tortuosity of pore spaces and porosity using the Kozeny–Carman equation [13]. Furthermore, pore size distribution and pore connectivity are conventional parameters that control hydrologic properties. Elastic properties of rock have often been formulated by crack-based models [11, 17, 20] or cemented grain models [3] given that grain connectivity is an important parameter with which to estimate elastic properties.

Recent developments in X-ray microcomputed tomography enable the pore geometry to be obtained in geological materials at micrometer resolution in 3D (Fig. 1). In our previous studies, we have used such images and applied numerical simulations to the digital rock models to directly estimate the hydraulic and elastic properties (Fig. 2a) [9, 21, 22]. The lattice Boltzmann method (LBM) was used to calculate the hydraulic properties of the digital rock model [7, 9, 14, 21]. LBM is advantageous with respect to dealing with complex boundaries and incorporating microscopic interactions in multi-physics. Using this numerical method, we calculated the behavior of CO2 in water-saturated rock for CO2 geological storage under various conditions, and the amount of CO2 stored in the natural rock pores was investigated (Fig. 3a; [21]). The background color in right panel of Fig. 3a indicates CO2 saturation under

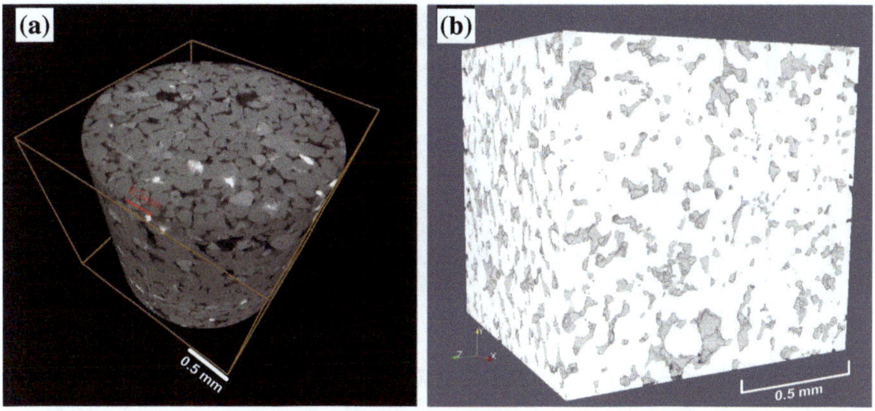

Fig. 1 Heterogeneous pore geometry of natural rock obtained by 3D computerized tomography (CT). **a** 3D rock model of Berea sandstone, reconstructed from a micro-CT scan [9]. White colors indicate higher CT values which indicate high-density mineral, while black colors indicate lower densities, such as pore spaces. **b** Binary model of pore geometry of Berea sandstone [21]. Persistent homology was applied to the digitized pore space geometry obtained from these scans

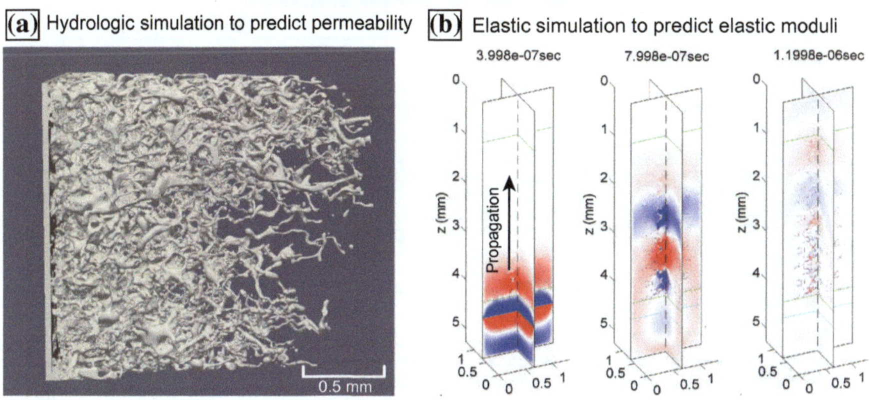

Fig. 2 Numerical approach to estimating hydrologic and elastic properties of the digital rock shown in Fig. 1. **a** Example of CO2 behavior in water-saturated digital rock model derived from the lattice Boltzmann simulation [21]. In this figure, the solid rock and water are transparent. **b** Wave propagation derived from dynamic wave propagation simulation [22] from which seismic velocity was calculated. Red and blue indicate the strain of the rock

several reservoir conditions (Lenormand diagram between capillary number and the viscosity ratio of two-phase flow [12]). It is found that saturation of CO2 can be estimated with two dimensionless parameters, namely capillary number (Ca) and viscosity ratio (M) [21]. Ca is represented by surface tension and viscous forces. Thus, these two dimensionless parameters are significant parameters with which to identify conditions of efficient CO2 storage. However, when the same approach is applied to different rock types, such as a homogeneous pore model, it is found

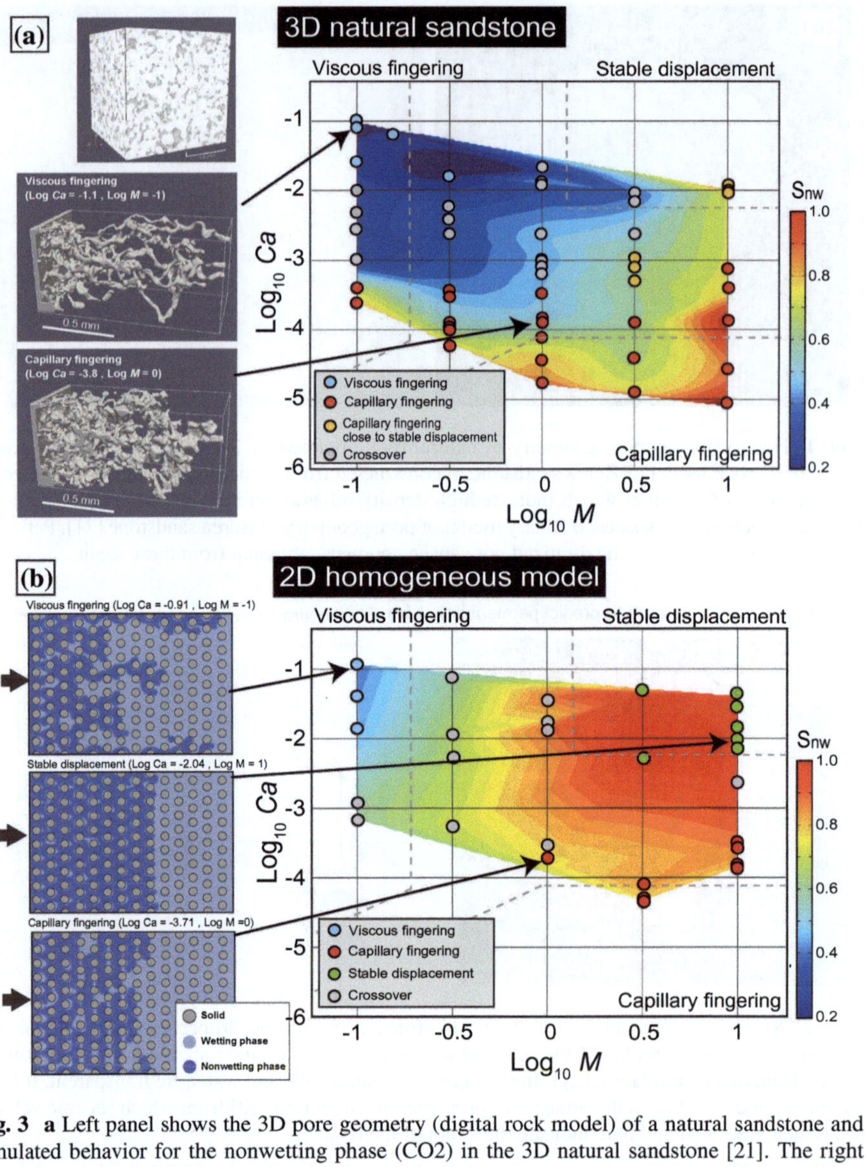

Fig. 3 **a** Left panel shows the 3D pore geometry (digital rock model) of a natural sandstone and simulated behavior for the nonwetting phase (CO2) in the 3D natural sandstone [21]. The right panel shows plots of CO2 saturation (background color) under various conditions of flow capillary number (*Ca*) and viscosity ratios (*M*). The CO2 infiltration behavior can be classified as primarily capillary fingering (red dots) or viscous fingering (blue dots), with little stable displacement. The gray dots indicate a crossover state that cannot be classified as one of the three typical displacement patterns. **b** Left panel shows nonwetting phase behavior in a 2D homogeneous pore model under various flow conditions, while the right panels show plots of the nonwetting phase (CO2) saturation for the 2D homogeneous model

that CO_2 behavior and CO_2 saturation are largely different (background color in Fig. 3b). This suggests that the rock pore geometry strongly influences fluid behavior and is important when modeling rock pore geometry to predict fluid behavior in geo-engineering. However, there is no established way to systematically describe heterogeneous rock pore space. The heterogeneity in geological formations is typically very different from artificial materials. Thus, exploring the complex spatial distribution of pore to characterize heterogeneity, and to relate this information to hydrological and elastic properties, is a demanding task.

By considering fluid behavior within pore space (Fig. 2a), we further numerically modeled mineralization process (i.e., rock evolution process; Fig. 4) [8]. Precipitation of minerals in geological materials is a fundamental process in the evolution of rocks, and rock strengthening associated with mineralization is also related to the deformation features of geologic sequences [10, 19]. However, the computational time of numerically simulating mineralization by considering porous flow is usually

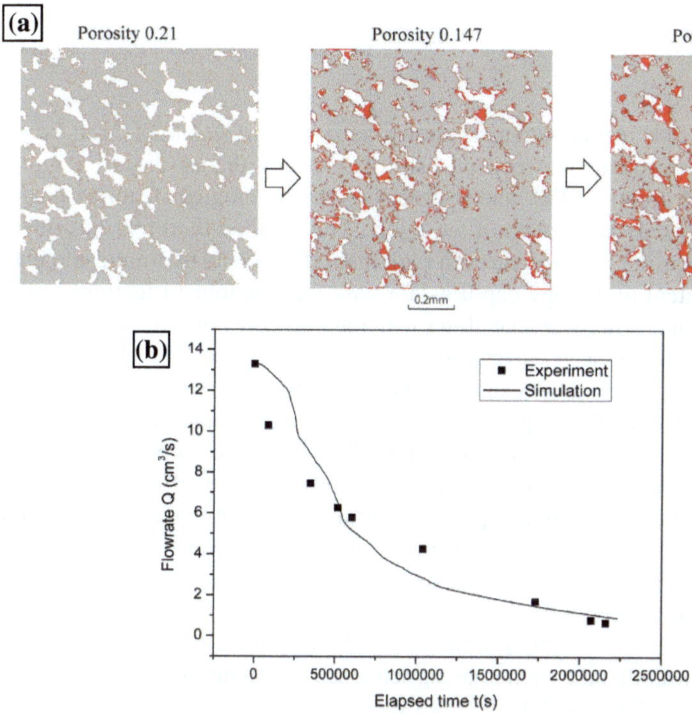

Fig. 4 Images of the rock evolution process generated using a numerical approach [8]. **a** Mineral precipitation calculated from CO_2 flow within the rock. Red represents precipitated calcite and white represents pore spaces. **b** Plot showing time variation of fluid flow rate (or permeability) which shows that flow rate decreases because of mineral precipitation and related decrease in permeability. The numerical results (solid line) are quite consistent with the laboratory-derived data (square dots) [23]

long, and the model size relatively restricted (mm to cm range). As a result, the model presented in Fig. 4a was run for a month to evaluate the evolution of mineral precipitation by considering porous flow [8] and we sought a practical approach to model in large-scale using a mathematical approach which allows for shorter computation time.

In present study, we applied a persistent homology for describing heterogeneous rock pore. By linking the diagram derived from persistent homology to the hydraulic and elastic properties, we attempt to develop a method for estimating these properties directly from the diagram. The rock evolution processes are further modeled using random-walk approach. By defining the probability and degree of mineral precipitation and diffusion in gridded cells, it is possible to model the time-variant pore geometry caused by mineralization.

2 Persistent Homology for Rock Pore Modeling

2.1 Methods

The recent approach in mathematics known as *persistent homology* can be used to characterize heterogeneous rock porosity. Homology theory dates back to the work of Poincaré around 1900, and it describes the topological features (such as connected components and holes) of an object. This method therefore has obvious application in revealing key structures in rock that influence its hydraulic and elastic properties. Persistent homology captures not only the topological features but also their appearance and disappearance along with parameters in the so-called filtration, which is used to convert the given data set into an increasing family of global objects. In the context of topological data analysis, persistent homology has been applied to, amongst others, data analysis, protein analysis, material sciences, and image data analysis. However, its application to natural rocks has not been previously reported.

A data set is often provided (or can be considered) as a point cloud data $\mathscr{X} \subset \mathbb{R}^d$ in the Euclidean space. The first step of computing persistent homology is to create a filtration of simplicial complexes from $\mathscr{X}$ in a certain way. Here, we explain the Vietoris–Rips complex of $\mathscr{X}$ as an example. A subset $\{v_0, v_1, \ldots, v_k\} \subset \mathscr{X}$ is a k-simplex in the Vietoris–Rips complex $\mathrm{VR}(\mathscr{X}, r)$ if $B_r(v_p) \cap B_r(v_q) \neq \emptyset$ for any $0 \leq p < q \leq k$, where $B_r(x)$ is the closed ball centered at x with radius r. In applications, we usually take an increasing sequence of real numbers $r_0 < r_1 < r_2 < \cdots$ and consider the filtration $\mathrm{VR}(\mathscr{X}) = \{\mathrm{VR}(\mathscr{X}, r_s)\}_{s \in \mathbb{Z}_{\geq 0}}$, where $\mathbb{Z}_{\geq 0} = \mathbb{N} \cup \{0\}$.

Let $\mathscr{K} = \{K_s\}_{s \in \mathbb{Z}_{\geq 0}}$ be a filtration of simplicial complexes, i.e., $K_s \subset K_t$ for $s \leq t$. We suppose that there exists $s_0 \in \mathbb{Z}_{\geq 0}$ such that $K_s = K_{s_0}$ for all $s \geq s_0$. The usual definition of the q-th homology of a simplicial complex K with coefficient field $\mathbb{F}$ provides a vector space $H_q(K)$. For $s \leq t$, we denote the linear map on homologies induced from the inclusion $K_s \hookrightarrow K_t$ by $\iota_t^s : H_q(K_s) \to H_q(K_t)$. The *$q$-th persistent homology* $H_q(\mathscr{K}) = (H_q(K_s), \iota_t^s)$ of $\mathscr{K}$ is defined by the family of homologies

$\{H_q(K_s)\}_{s\in\mathbb{Z}_{\geq 0}}$ and the induced linear maps ι_t^s for all $s \leq t$. Here, we assume that $\dim H_q(K_s) < \infty$ for every $s \in \mathbb{Z}_{\geq 0}$. In this setting, a persistent homology $H_q(\mathscr{K})$ has a nice decomposition property.

Theorem 1 ([24]) *There uniquely exist $M \in \mathbb{Z}_{\geq 0}$ and $b_m, d_m \in \overline{\mathbb{Z}}_{\geq 0} = \mathbb{Z}_{\geq 0} \sqcup \{\infty\}$ with $b_m < d_m$, $m = 1, 2, \dots, M$ such that the following isomorphism holds:*

$$H_q(\mathscr{K}) \cong \bigoplus_{m=1}^{M} I(b_m, d_m). \tag{2}$$

Here, $I(b_m, d_m) = (V_s, f_t^s)$ consists of a family of $\mathbb{F}$-vector spaces

$$V_s = \begin{cases} \mathbb{F}, & b_m \leq s < d_m, \\ 0, & \text{otherwise}, \end{cases}$$

and the identity map $f_t^s = \mathrm{id}_{\mathbb{F}}$ for $b_m \leq s \leq t < d_m$.

Each summand $I(b_m, d_m)$ in (2) is called a generator of the persistent homology and (b_m, d_m) is called its birth–death pair, which represents the appearance and disappearance of the i-th topological feature. From the uniqueness of the decomposition, the multiset of birth–death pairs $D_q(\mathscr{K}) = \{(b_m, d_m) \in \overline{\mathbb{Z}}^2_{\geq 0} : m = 1, \dots, M\}$ can be viewed as the characteristics of a given filtration in persistent homology and is called the q-th persistence diagram. In applications, we usually take an increasing sequence of parameters $\{r_s\}_{s\in\mathbb{Z}_{\geq 0}}$ and identify $s \in \mathbb{Z}_{\geq 0}$ with $r_s \in \overline{\mathbb{R}}_{\geq 0}$ to obtain

$$D_q(\mathscr{K}) = \{(r_{b_m}, r_{d_m}) \in \overline{\mathbb{R}}^2_{\geq 0} : m = 1, \dots, M\}.$$

It can be visualized in a compact form as displayed in Figs. 5 and 6.

The calculation of a persistence diagram for complex pore geometry can be divided into three parts. First, the images of rock samples are obtained by a multislice X-ray microtomography scanner (Fig. 1a). A segmentation process is then performed to delineate pores and minerals in the images to reconstruct the binary 3D model (Fig. 1b). Then, a cloud of spheres which represent the spatial distribution of pore bodies is obtained by a maximal ball algorithm [2]. In this approach, the maximal ball algorithm constructs the largest spheres centered at each void voxel (a volume of 3D space) that just fits in the pore space. A maximal ball is one such sphere that is not completely enclosed by another. Finally, the analysis of persistent homology is carried out from the extracted sphere cloud data (Figs. 5 and 6). Therefore, the persistence diagram we estimated in this study (Figs. 5 and 6) reflects geometry of central position of pore body and pore size. In this study, the open-source code "Perseus" [1] was used to calculate the Vietoris–Rips complex topology of a set of points and consequently determine the persistence diagram [15]. Perseus software is based on discrete Morse theory [6], and therefore does not rely on the idiosyncrasies of a particular type of complex structure or dimension for its efficiency. The persistent homology

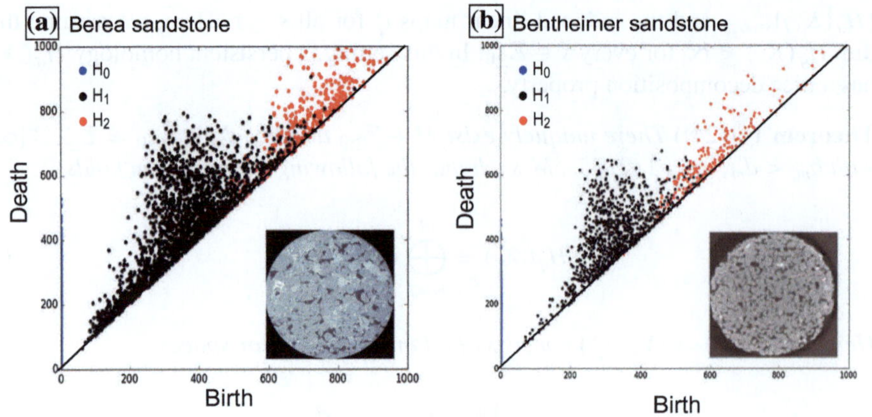

Fig. 5 Persistence diagrams of two natural sandstones. **a** Berea sandstone and **b** Bentheimer sandstone. The color of the dots in the BD plots indicates the different dimensions, i.e., H_0, H_1, and H_2, in which H_0 (blue dots) are aligned parallel to the vertical axis

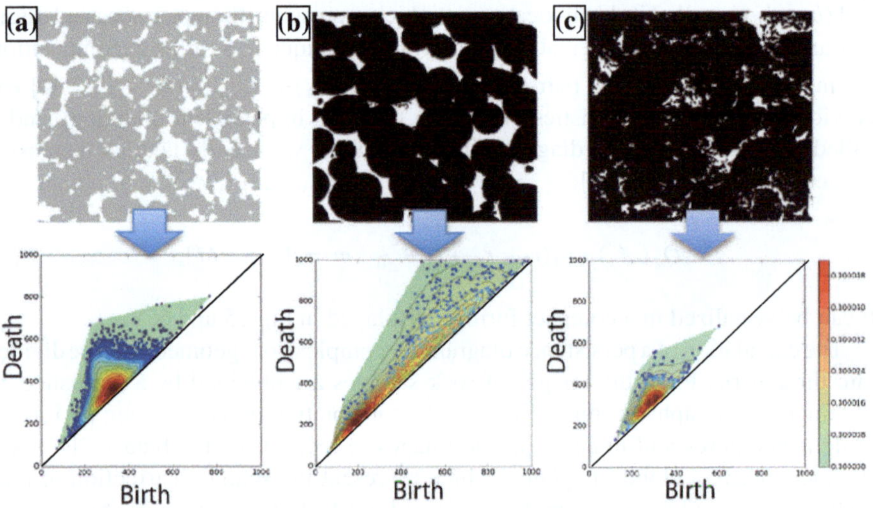

Fig. 6 Persistence diagrams of different rock types. The rock model in panel (**a**) is the same as that in Fig. 5a. The upper three panels show CT slice images of different rock types with their respective persistence diagrams shown in the lower panels. In the CT slices, the pore spaces are white in color. The persistence diagrams displayed are from H_1, in which the color indicates the density of the BD plots. The color bar indicates the contour level of the probability density function for each point in persistence diagram, which is estimated by the nonparametric Gaussian kernel density estimation (KDE). Single BD points are highlighted by dots

was then calculated for Vietoris complexes generated around three different types of data: uniform birth points, nonuniform birth points, and distance matrices. The most common type of Vietoris–Rips complexes are uniform birth points. The input data consist of a list of vertices (points) embedded in a Euclidean space. For each vertex, an initial radius r is obtained from the data of sphere size distributions. The radius for each vertex is incrementally applied N times by a universal step size to give an increasing sequence of radii for each point. In this study, the incremented time N was set to 1000, with an increment step size of 0.1 μm. Changing parameters, including initial conditions of the sphere in the analysis, would result in production of different diagrams.

2.2 Results and Discussions

Figure 5 shows examples of persistence diagrams for H_0, H_1, and H_2 derived from this analysis. It is evident that the similarity in pore geometries of the two sandstones displayed in Fig. 5 results in the two model diagrams having similar characteristics. The birth–death (BD) points of H_0 are located on the vertical axis. The BD points of H_1 and H_2 are different: the BD points of H_1 are far from the diagonal line, and the birth time of H_1 is earlier than that of H_2. In the following discussions, the H_1 diagram is characterized.

Persistent homology was then applied to three rock types with different characteristics (Fig. 6), from which it is evident that the features of the persistence diagrams are largely different for different rock types. These differences originate because of differences in pore geometry (e.g., relative location of pore body) and may also reflect the hydraulic and elastic properties. Several BD points of the homogeneous sandstone (homogeneously distributed pore body; blue dots in Fig. 6a) are far from the diagonal line in the persistence diagram. The longer BD cycle evident from the distance of the points from the diagonal line indicates that pore network has large circle structure. It is observed that local concentration of separated pore bodies leads to a short life cycle because the topological rings disappear soon after their formation. The wide range of birth time of the rock model in Fig. 6b indicates heterogeneous characteristics of the rock because many kinds of H_1 cycles are generated in the filtration.

As discussed, the persistence diagrams reflect the topological features of rock pore. Thus, the persistence diagrams could be used to estimate the hydraulic properties of rock because they are influenced by pore connectivity. Additionally, if persistent homology is applied to the grains (solid particles), it is possible to characterize the elastic properties which are largely influenced by grain connectivity.

3 Random Walk in Rock Mineralization Modeling

3.1 Methods

We applied a random-walk particle method (stochastic method) to simulate CO2 transport and CO2 mineralization in porous rocks. A random walk is a trajectory comprised of a series of random steps. In general, random-walk method requires little computer storage and does not suffer from numerical dispersion in problems dominated by advection. Therefore, the method is often used for simulating transport in large and heterogeneous flow systems [18].

Particles move on a regular lattice to neighboring lattice nodes through a path between the two nodes according to some probability distribution. In our simulation, the probability distribution is determined by magnitudes of fluxes due to each transport process, such as advection and diffusion. We gave the values of the flow velocity (v) for advection and the diffusivity (D) for diffusion on each path of the lattice grids. By using spatial distribution of the flow velocity or the diffusivity, this random-walk simulation can take spatial distribution of the transport processes into account. Nevertheless, for simplicity, this work shows that the flow area is homogeneous and isotropic and that the grain and the pore space are not distinguished in the initial condition. Thus, we set constant values of v and D to 1.0 and 0.01, respectively, on each path on the regular lattice grids for the initial condition. Although we used certain values for these parameters in this study, reasonable values should be obtained from laboratory experiments or field observations. The migration times for molecules passing through the paths on the lattice were determined by using v and D for advective and diffusive processes. The migration times of each process were converted to the flux in each direction, respectively. Each probability of the transport processes in each direction is determined by the ratio of the magnitude of the flux to the total flux.

We used two-dimensional domain. A group of CO2 molecules were injected from the start points (left side in Fig. 7) and traveled until arriving at goals (right side in Fig. 7) at each time step. We used a hundred CO2 molecules in a group at each time step. When a hundred molecules arrived at the goal, the next group was injected at the next time step. One-dimensional advective flow is given from the left to the right in the calculation domain. The diffusion occurs into all directions. Note that the upper and lower boundaries were periodic condition.

The mineralization process was also considered. It is assumed that the more CO2 particles passed, the easier it is to mineralize CO2 at the location. When mineralization occurs at a lattice path, the path is closed, and molecules are no longer able to move through the closed path. We counted the number of times that particles passed through each path (k). The probability of mineralization p_M was defined as

$$p_M = \sum_{i=1}^{k} (1-q)^{i-1} q, \tag{3}$$

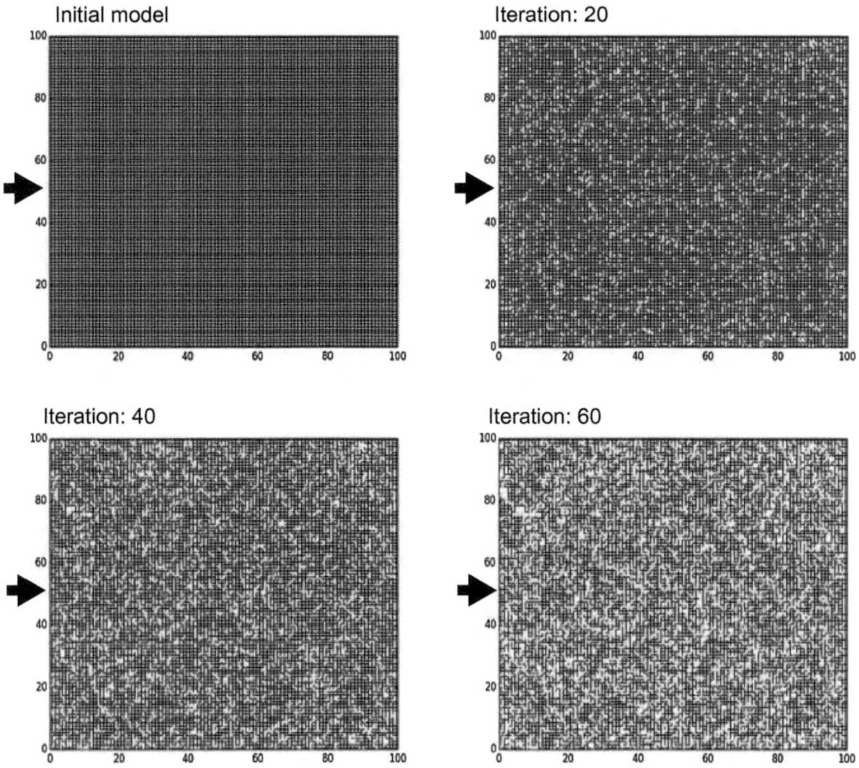

Fig. 7 Visual representation of how random walk is used to model rock evolution, in this case mineral precipitation. Black arrows indicate the fluid inlet (left) into the model. Black lines are open bonds and white lines are closed bonds caused by mineralization. These processes have similar meaning with Fig. 4a. The mineralization parameter q is 0.01 in this modeling

where q is the mineralization parameter. The larger number of times that particles passed, the higher probability of mineralization p_M is. The parameter q may be able to be determined based on laboratory experiments or field observations.

3.2 Results and Discussions

Temporal-spatial variation of mineralization on the lattice grid simulated by the random-walk model is shown in Fig. 7. Number of grid cells in this modeling is 100×100. All paths opened for the initial condition. The black lines describe the open paths. When mineralization occurs, the path is closed, which is described by the white lines in Fig. 7. The results shown in Fig. 7 are similar with the mineralization by reactive transport modeling, as shown in Fig. 4a. In this example, we used the mineralization parameter of 0.01, which is for emphasizing mineralization process.

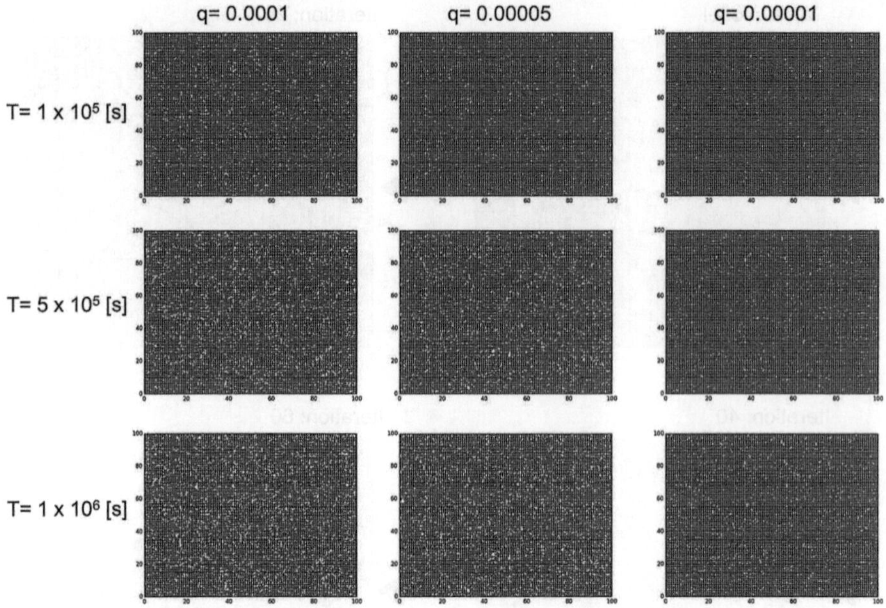

Fig. 8 Mineralization process derived from random-walk modeling. Here, we used three mineralization parameters ($q = 0.0001$, 0.00005, and 0.00001). These figures are similar to Fig. 7, but the speed of mineralization is slow (natural)

The effect of mineralization parameters q is shown in Fig. 8. We varied the values of q and evaluated the influence upon their mineral precipitation processes by comparing with experimental result [8] (Fig. 4). We used the mineralization parameter $q = 0.0001, 0.00005$, and 0.00001 in Fig. 8. Note that the elapsed time in Fig. 8 was given by

$$\text{Elapsed time} = \sum_{i=1}^{N} T_i, \tag{4}$$

where N is the number of iteration and T_i is the average travel time of injected particles in the i-th time step. As shown in Fig. 8, the mineralization features, such as mineralization speed, depends on the value of q.

We evaluated the influence of the precipitation on hydraulic properties, such as permeability and porosity. Permeability is proportional to the flow rate Eq. (1), that is, the reciprocal of travel time from the inlet to the outlet. The travel time for each molecule was calculated by accumulating the migration times of passing through the lattice paths. Figure 9a presents the time variations of permeability with a different precipitation parameter q. The permeability was normalized by the travel time that molecule moves due to advection between the inlet and the outlet linearly. The white circles in Fig. 9a are the permeability decline derived from the laboratory experiment

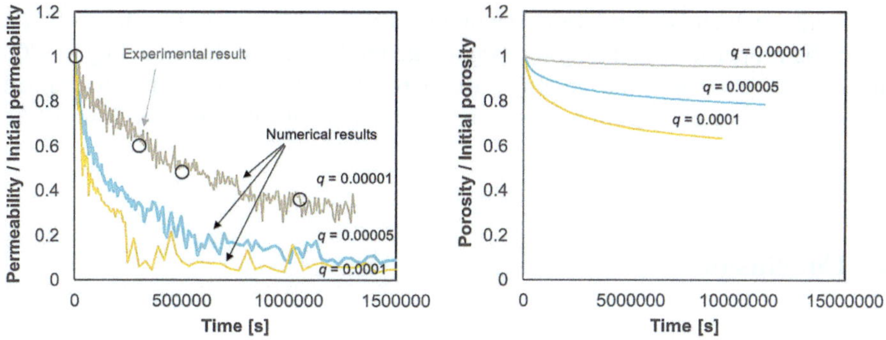

Fig. 9 Time variation of hydrological properties derived from random walk modeling. Here, we display the results obtained at 3 different mineralization parameters ($q = 0.0001, 0.00005, and\ 0.00001$). **a** Permeability reduction caused by mineral precipitation according to the random-walk model. In this panel, we display the results of laboratory experiments (circles) [23]. **b** Variation in porosity caused by mineralization according to the random-walk model

[23] shown in Fig. 4b. As shown in Fig. 9a, the simulation results of permeability decreased with time, which is consistent with the experimental result. The simulation result with $q = 0.00001$ was closer to the experimental results.

Porosity is also expected to decrease because of mineral precipitation. The porosity (Φ) is given by the following equation:

$$\Phi(t) = 1 - \frac{B_c(t)}{B_{total}}, \tag{5}$$

where t is the elapsed time, B_c is the number of closed paths, and B_{total} is the total paths. Figure 9b depicts the variation of porosity. In the simulation, the porosity with $q = 0.00001$ decreased to 70% of the initial porosity at 3.8×10^6 s (after 3660 time steps). The result indicates that the random-walk model would provide a simulation of temporal-spatial variation of hydraulic properties (e.g., permeability and porosity) due to the mineral precipitation.

We will validate the random-walk simulations for CO2 mineralization. Further research will use the image of rock grains obtained from multi-slice X-ray microtomography (Fig. 4) as an initial rock condition. The images of rock grains can provide information that each path is opened or closed with the same shapes as porous rocks. In this study, we introduced the precipitation parameters q. As shown in Fig. 9, the precipitation parameter affects the variation of hydraulic properties. Spatial validation of the precipitation parameter may be effective to conduct more detailed simulation for mineralization in pore space. For instance, because the probability of mineralization could be higher at interface between grain and pore, the mineralization probability can be increased at the interface. The simulation parameters such as velocity (v), diffusivity (D), and mineralization parameter (q) should be validated by features of natural rock or laboratory experiment. Nevertheless, this random-walk approach can provide relatively short simulation time and flexible treatment of mineralization parameters, which is attractive to obtain temporal-spatial variation in macroscopic hydraulic properties (modeling for large-scale model). Thus, the modeling of CO2 mineralization based on the random-walk approach can be a practical tool to evaluate the influence of CO2 injection and to understand mineralization processes in porous rocks.

4 Conclusions

The hydraulic and elastic properties of natural rocks are largely controlled by their pore geometry, but modeling of heterogeneous rock pore has not been well established in the literature. In this study, pore geometry of natural rocks was modeled using persistent homology. The characteristics of persistence diagrams derived from this analysis are different for each rock type, and thus appear to reflect differences in rock pore geometry. By linking the persistence diagrams of natural rocks with their hydraulic and elastic properties, we attempt to develop a method for estimating these properties directly from the persistence diagrams.

Rock evolution processes, such as mineral precipitation, were also modeled using random-walk simulation. The mineral precipitation modeling based on pore fluid flow is time-consuming, and thus long-term mineralization modeling in pore space ($\sim$100 years) is not realistic within the present computational environment. The calculation time using random walk is much shorter and, based on the results of random-walk modeling of precipitated mineral distribution, the temporal variation in permeability and porosity can be predicted, as well as providing useful hydraulic properties. This approach is flexible and can essentially model any mineralization processes. Parameters used in the random-walk modeling can be determined by comparison with the temporal variation in hydraulic properties derived from laboratory experiments or by comparison with mineral distribution in natural rocks.

Acknowledgements This research was improved by discussions in the Study Group Workshop in 2016 and is supported by a joint project between the International Institute for Carbon-Neutral Research (I2CNER) and Institute of Mathematics for Industry (IMI), Kyushu University. This work was partially supported by JSPS through a Grant-in-Aid for Science Research on Innovative Area (no.JP15H01143; JP17H05318). T.S. is partially supported by JSPS Grant-in-Aid (26610025, 26287019) and JST CREST Mathematics (15656429).

References

1. Cubical complex of Perseus software project Web page, http://www.sas.upenn.edu/vnanda/perseus/index.html
2. H. Dong, M.J. Blunt, Pore-network extraction from micro-computerized-tomography images. Phys. Rev. E **80**(3), 36307 (2009), https://doi.org/10.1103/PhysRevE.80.036307
3. J. Dvorkin, A. Nur, H. Yin, Effective properties of cemented granular material. Mech. Mater. **18**, 351–366 (1994)

4. H. Edelsbrunner, J. Harer: Persistent homology—a survey, in *Surveys on Discrete and Computational Geometry*. Contemp. Math., Vol. 453 (Amer. Math. Soc., Providence, 2008), pp. 257–282
5. R.E. Ewing,*The Mathematics of Reservoir Simulation*, (SIAM, 1983)
6. R. Forman, Morse theory for cell complexes. Adv. Math. (N.Y) **134**(1), 90–145 (1998), https://doi.org/10.1006/aima.1997.1650
7. H. Huang, L. Wang, X.Y. Lu, Evaluation of three lattice Boltzmann models for multiphase flows in porous media. Comput. Math. Appl. **61**(12), 3606–17 (2011), https://doi.org/10.1016/j.camwa.2010.06.034
8. F. Jiang, T. Tsuji, Changes in pore geometry and relative permeability caused by carbonate precipitation in porous media. Phys. Rev. E **90**, 053306 (2014), https://doi.org/10.1103/PhysRevE.90.053306
9. F. Jiang, T. Tsuji, Estimation of three-phase relative permeability by simulating fluid dynamics directly on rock-microstructure images. Water Resour. Res. (2017), https://doi.org/10.1002/2016WR019098
10. H. Kopp, N. Kukowski, Backstop geometry and accretionary mechanics of the Sunda margin. Tectonics **22**(6), 1072 (2003), https://doi.org/10.1029/2002TC001420
11. G.T. Kuster, M.N. Toksoz, Velocity and attenuation of seismic waves in two-phase media, part 1 theoretical formulations. Geophysics **39**(5), 587–606 (1974), https://doi.org/10.1190/1.144050
12. R. Lenormand, E. Touboul, C. Zarcone, Numerical models and experiments on immiscible displacements in porous media. J. Fluid Mech. **189**(9), 165–187 (1988)
13. G. Mavko, A. Nur, The effect of a percolation threshold in the Kozeny-Carman relation. Geophysics **62**(5), 1480–1482 (1997), https://doi.org/10.1190/1.1444251
14. G. McNamara, G. Zanetti, Use of the Boltzmann equation to simulate lattice-gas automata. Phys. Rev. Lett. **61**, 2332 (1988), https://doi.org/10.1103/PhysRevLett.61.2332
15. K. Mischaikow, V. Nanda, Morse theory for filtrations and efficient computation of persistent homology. Discret. Comput. Geometr. **50**(2), 330–353 (2013)
16. PHAT (Persistent Homology Algorithm Toolbox), https://code.google.com/p/phat/
17. P.M. Shearer, Cracked media, Poisson's ratio and the structure of the upper oceanic crust. Geophys. J **92**, 357–362 (1988)
18. A.F.B. Tompson, L.W. Gelhar, Numerical simulation of solute transport in three-dimensional, randomly heterogeneous porous media. Water Resour. Res. **26**(10), 2541–2562 (1990)
19. T. Tsuji, J. Ashi, M. Strasser, G. Kimura, Identification of the static backstop and its influence on the evolution of the accretionary prism in the Nankai Trough. Earth Planet. Sci. Lett. **431**, 15–25 (2015), https://doi.org/10.1016/j.epsl.2015.09.011
20. T. Tsuji, G.J. Iturrino, Velocity-porosity relationships of oceanic basalt from eastern flank of the Juan de Fuca ridge: the effect of crack closure on seismic velocity. Explor. Geophys. **39**(1), 41–51 (2008), https://doi.org/10.1071/EG08001
21. T. Tsuji, F. Jiang, K. Christensen, Characterization of immiscible fluid displacement processes with various capillary numbers and viscosity ratios in 3D natural sandstone. Adv. Water Resour. **95**, 3–15 (2016), https://doi.org/10.1016/j.advwatres.2016.03.005
22. H. Yamabe, T. Tsuji, Y. Liang, T. Matsuoka, Influence of fluid displacement patterns on seismic velocity during supercritical CO2 injection: simulation study for evaluation of the relationship between seismic velocity and CO2 saturation. Int. J. Greenh. Gas Control **46**, 197 204 (2016), https://doi.org/10.1016/j.ijggc.2016.01.011
23. S.-Y. Yoo, Y. Kuroda, Y. Mito, T. Matsuoka, M. Nakagawa, A. Ozawa, K. Sugiyama, A. Ueda, A geochemical clogging model with carbonate precipitation rates under hydrothermal conditions. Appl. Geochem. **30**, 67–74 (2013)
24. A. Zomorodian, G. Carlsson, Computing persistent homology. Discrete Comput. Geom. **33**, 249–274 (2005)

Statistical Machine Learning for Agriculture and Human Health Care Based on Biomedical Big Data

Yoshihiro Yamanishi, Yasuo Tabei and Masaaki Kotera

Abstract The availability of biomedical big data provides an opportunity to develop data-driven approaches in agriculture and human healthcare research. In this study, we investigate statistical machine learning approaches to metabolic pathway reconstruction and the prediction of drug–target interactions, using heterogeneous biomedical big data. We present an L_1-regularized pairwise support vector machine to predict unknown enzymatic reactions among metabolome-scale compounds, based on chemical transformation patterns of compounds. We also present supervised bipartite graph inference with kernel methods to predict unknown interactions between drugs and target proteins, based on the chemical structures of drugs and the amino acid sequences of proteins. We experimentally demonstrated that these methods could be applied to rational compound synthesis and efficient drug discovery for a range of human diseases. Such methods are expected to increase the productivity of research in food and pharmaceutical industries.

Keywords Metabolic pathways · Drug targets · Machine learning
Classification · Feature extraction · Graph inference

Y. Yamanishi (✉)
Division of System Cohort, Medical Institute of Bioregulation, Kyushu University, 3-1-1 Maidashi, Higashi-ku, Fukuoka 812-8582, Japan
e-mail: yamanishi@bioreg.kyushu-u.ac.jp

Y. Yamanishi
Institute for Advanced Study, Kyushu University, 6-10-1, Hakozaki, Higashi-ku, Fukuoka 812-8581, Japan

Y. Yamanishi
PRESTO, Japan Science and Technology Agency, Kawaguchi, Saitama 332-0012, Japan

Y. Tabei
RIKEN Center for Advanced Intelligence Project, Nihonbashi 1-chome Mitsui Building, 15th foor, 1-4-1 Nihonbashi, Chuo-ku, Tokyo 103-0027, Japan
e-mail: yasuo.tabei@riken.jp

M. Kotera
School of Life Science and Technology, Tokyo Institute of Technology, 2-12-1 Ookayama, Meguro-ku, Tokyo 152-8550, Japan
e-mail: maskot@bio.titech.ac.jp

R. S. Anderssen et al. (eds.), *Agriculture as a Metaphor for Creativity in All Human Endeavors*, Mathematics for Industry 28,
https://doi.org/10.1007/978-981-10-7811-8_12

1 Introduction

Recent advances in biomedical science have made a wide range of omics data (e.g., genome, transcriptome, proteome, and metabolome) available, providing a resource for data-driven approaches in agriculture and human health care. Omics-based analysis of metabolic pathways is important in a range of applications [1]. Knowledge of the plant metabolism (e.g., metabolites, reactions, and enzymes) can be used to support rational cultivation in agriculture. Enzymes can be therapeutic targets of many human diseases, and metabolites (natural products) have long been used as drugs or drug leads in the pharmaceutical sector. Approximately, 50% of all anticancer drugs are derived from compounds produced by microbes, fungi, or plants [2], and some human metabolites are now providing effective biomarkers for the diagnosis of certain diseases [3].

It is estimated that more than one million compounds exist in the plant kingdom [4], but the reaction steps and the associated enzymes for most compounds in the metabolic pathways have not yet been identified [5]. Experimental identification of novel enzymatic reactions is extremely challenging, and there is a need to develop computational methods for reconstructing metabolic pathways that are not yet understood [6]. Some such computational methods predict potential reaction steps by hypothesizing the intermediate compounds between the source and target [7–12]. Other methods predict the enzymatic reactions among compounds from chemical interconversion [13–18].

Knowledge of the metabolic pathways can also support the discovery of new drugs [19]. Most drug molecules interact with target proteins (including enzymes) and modulate the associated biological pathways (including the metabolic pathways). The identification of drug–target interactions, interactions between drugs (or candidate compounds) and target candidate proteins, is, therefore, a crucial stage of drug development. Many machine learning approaches have been developed for the prediction of drug–target interactions, based on chemical data of drugs and genomic data of target proteins [20]. Existing methods can be categorized into those using binary classification [21–25] and those using dimension reduction [26–29].

In this study, we introduce statistical machine learning methods that allow reconstruction of metabolic pathways and prediction of drug–target interactions, using heterogeneous biomedical big data. We present an L_1-regularized pairwise support vector machine to predict unknown enzymatic reactions among metabolome-scale compounds, based on chemical transformation patterns of compounds [16]. We also present supervised bipartite graph inference with kernel methods to predict unknown interactions between drugs and target proteins, based on chemical structures of drugs and amino acid sequences of proteins [26, 27]. Our experiments demonstrated the usefulness of these methods for rationalizing compound synthesis and improving drug discovery. This paper discusses the characteristics of each method and suggests directions for future research.

2 Methods for Metabolic Pathway Reconstruction

2.1 *Formalism*

Metabolic pathways comprise a series of enzymatic reactions among metabolites (chemical compounds), and reconstruction can therefore be treated as the problem of estimating the enzymatic-reaction likeness of any given compound–compound pair, and predicting whether each compound in the pair can be converted to the other by an enzymatic reaction [16]. Figure 1 sets out an illustration of the problem, with circles indicating compounds and rectangles indicating enzymatic reactions. Our approach to the estimation of the enzymatic-reaction likeness of each compound–compound pair was conducted within a supervised classification framework.

Suppose that we have a set of compounds $\{\mathbf{x}_i\}_{i=1}^n$, where n is the number of compounds. Given a set of $n(n-1)$ compound–compound pairs $(\mathbf{x}_i, \mathbf{x}_j)(i = 1, ..., n, j = 1, ..., n, i \neq j)$, we estimate a function $f(\mathbf{x}, \mathbf{x}')$ that predicts whether or not compound $\mathbf{x}$ is converted to $\mathbf{x}'$ in an enzymatic reaction. The key biochemical features of the enzymatic reactions should also be identified. In practical applications, compound–compound pairs in known enzymatic reactions (substrate–product pairs) are used as positive examples. Other compound–compound pairs are used as negative examples, because it is not possible to obtain true negative data for enzymatic reactions.

2.2 *Predictive Model*

We consider the use of linear models for predicting the enzymatic-reaction likeness, as these are used in many statistical tasks involving high-dimensional data, such as classification and regression. The models also have good interpretability.

Let $\mathbf{x}$ and $\mathbf{x}'$ be two chemical compounds. Compounds $\mathbf{x}$ and $\mathbf{x}'$ are represented by their d-dimensional fingerprints (binary vectors) as $\Phi(\mathbf{x}) = (x_1, x_2, ..., x_d)^T$

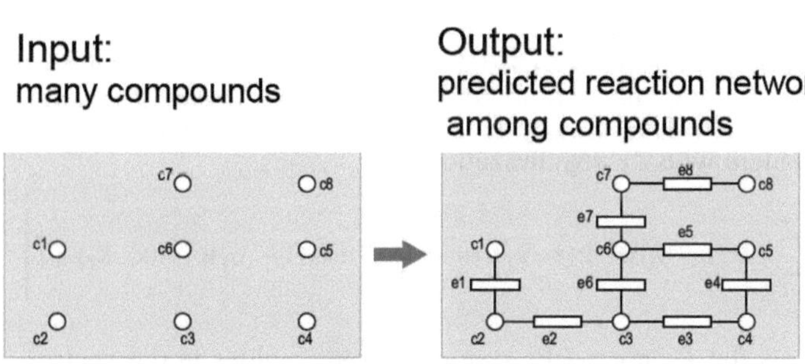

Fig. 1 The process of metabolic pathway reconstruction

and $\Phi(\mathbf{x}') = (x'_1, x'_2, ..., x'_d)^T$, respectively, where $x_k, x'_k \in \{0, 1\}$, $k = 1, ..., d$. To make use of existing machine learning techniques, we represent each compound–compound pair by a feature vector $\Phi(\mathbf{x}, \mathbf{x}')$ and estimate a function $f(\mathbf{x}, \mathbf{x}') = \mathbf{w}^T \Phi(\mathbf{x}, \mathbf{x}')$. The design of feature vectors is fundamental to both predictive power of the model and its interpretability. We construct a high-dimensional feature vector for each compound–compound pair, based on structural changes in the two compounds [16]. We define two fingerprinting operations as follows:

$$(\Phi(\mathbf{x}) \wedge \Phi(\mathbf{x}')) = (I(x_1 = x'_1 = 1), ..., I(x_d = x'_d = 1)),$$

and

$$(\Phi(\mathbf{x}) \ominus \Phi(\mathbf{x}')) = (I(x_1 = 1, x'_1 = 0), ..., I(x_d = 1, x'_d = 0)),$$

where $I(\cdot)$ is an indicator function that returns a value of one if the event is true. The first operation is expected to capture any common features in $\Phi(\mathbf{x})$ and $\Phi(\mathbf{x}')$, and the second operation is expected to capture any features that are present in $\Phi(\mathbf{x})$ but absent from $\Phi(\mathbf{x}')$. This yields the following high-dimensional feature vector for each compound–compound pair:

$$\Phi(\mathbf{x}, \mathbf{x}') = (\Phi(\mathbf{x}) \wedge \Phi(\mathbf{x}'), \Phi(\mathbf{x}) \ominus \Phi(\mathbf{x}'), \Phi(\mathbf{x}') \ominus \Phi(\mathbf{x}))^T.$$

Note that the feature vector is asymmetrical: $\Phi(\mathbf{x}, \mathbf{x}') \neq \Phi(\mathbf{x}', \mathbf{x})$.

2.3 L_1-Regularized Pairwise Support Vector Machine (L1SVM)

We use linear support vector machine (SVM) as the binary classifier. Regularization is necessary to produce the generalization properties, especially when the feature vectors are high-dimensional. L_2-regularization is a commonly used approach, but this keeps most elements at nonzero values, making it difficult to interpret the model. An alternative approach is L_1-regularization, which forces most elements in the weight vector to take zero values. The resulting sparsity makes the model straightforward to interpret. For this reason, we used L_1-regularization in the linear SVM.

Given a set of compound–compound pairs and their labels $(\Phi(\mathbf{x}_i, \mathbf{x}_j), y_{ij})$ where $y_{ij} \in \{+1, -1\}$ $(i = 1, ..., n, j = 1, ..., n, i \neq j)$, we consider the following minimization problem with L_1-regularization:

$$\min_{\mathbf{w}} \sum_{i=1}^{n} \left\{ \sum_{j=1}^{i-1} \max\{1 - y_{ij}\mathbf{w}^T \Phi(\mathbf{x}_i, \mathbf{x}_j), 0\} + \sum_{j=i+1}^{n} \max\{1 - y_{ij}\mathbf{w}^T \Phi(\mathbf{x}_i, \mathbf{x}_j), 0\} \right\} + \lambda ||\mathbf{w}||_1, \quad (1)$$

where $||\cdot||_1$ is the L_1 norm (the sum of absolute values in the vector), and λ is a hyper-parameter [30]. L_1-regularized pairwise SVM is referred to as L1SVM, while L_2-regularized pairwise SVM is referred to as L2SVM.

The learning on a huge number of compound–compound pairs (the product of all compounds) using a conventional optimization algorithm is a nontrivial problem. As, in this context, $\Phi(\mathbf{x}, \mathbf{x}')$ is a sparse binary vector, it is possible to estimate weight vectors using efficient optimization algorithms [31] and min-wise hashing algorithms [32].

3 Methods for Drug–Target Interaction Prediction

3.1 Formalism

The drug–target interactions can be treated as a graph in which drugs (or candidate compounds) and target proteins (or candidates) are two different types of nodes and their interactions are edges [26, 27]. Mathematically, the drug–target interaction network is represented as the bipartite graph $G = (U + V, E)$, where $U = (\mathbf{x}_1, \ldots, \mathbf{x}_{n_x})$ is the set of drug nodes, $V = (\mathbf{z}_1, \ldots, \mathbf{z}_{n_z})$ is the set of target protein nodes, and $E \subset (U \times V)$ is the set of drug–target interaction edges.

We formulate the prediction of drug–target interactions as a problem of supervised bipartite graph inference with the goal of predicting potential edges between two different types of nodes. Figure 2 shows an illustration of the problem, where solid lines indicate known interactions and dot lines indicate previously unknown interactions (potentially true interactions to be predicted). The task is to predict interactions between any query drug $\mathbf{x}'$ and any query target protein $\mathbf{z}'$.

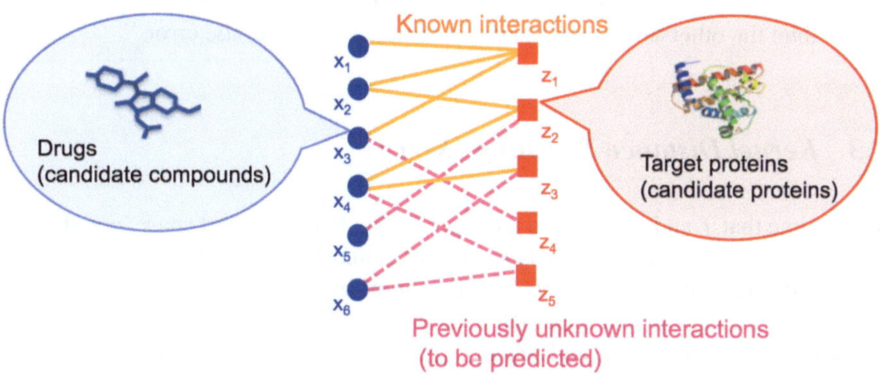

Fig. 2 The drug–target interaction prediction process

3.2 Euclidean Embedding

Here we introduce a machine learning method based on Euclidean embedding and distance learning [26, 27]. It involves two steps:

- Learn two functions f and g to embed the drugs and target proteins into a unified Euclidean space representing the bipartite graph structure in which interacting pairs of drugs and target proteins are closely co-located.
- Apply function f to drugs whose interaction partners are unknown and function g to target proteins whose interaction partners are unknown, and predict the interactions that will take place when the drugs and proteins are closely co-located.

The first step can be optimized by supervised learning of f and g, using the observed data and the partially known graph structure. f and g are expected to map adjacent nodes in the known bipartite graph onto nearby positions in a unified Euclidean space $\mathbf{R}^d$, ensuring that the nearest neighbor approach can recover the known bipartite graph structure.

We consider two functions $f : U \rightarrow \mathbf{R}$ and $g : V \rightarrow \mathbf{R}$ that map drugs and target proteins in a unified feature space, where the interacting drugs and target proteins are close to each other. To assess whether interacting drug–target pairs are mapped onto points that are close in $\mathbf{R}$, we apply the following minimization:

$$\min_{f,g} \frac{\sum_{(\mathbf{x}_i,\mathbf{z}_j)\in E}(f(\mathbf{x}_i) - g(\mathbf{z}_j))^2 - \sum_{(\mathbf{x}_i,\mathbf{z}_j)\notin E}(f(\mathbf{x}_i) - g(\mathbf{z}_j))^2}{\sum_{(\mathbf{x}_i,\mathbf{z}_j)\in U\times V}(f(\mathbf{x}_i) - g(\mathbf{z}_j))^2}, \tag{2}$$

where E is a set of known interaction edges on the bipartite graph. If the value of the objective function is small, this means that the interacting drug–target pairs are closer than the other drug–target pairs, in terms of the quadratic error.

3.3 Kernel Distance Learning (KDL)

We assume that f and g belong to the reproducing kernel Hilbert space, and that $\mathcal{H}_x$ and $\mathcal{H}_z$ are defined by the kernels k_x for drugs and k_z for target proteins. To avoid the overfitting problem, we apply regularization by a smoothness functional on f and g, based on a classical approach used in statistical learning [33, 34]. We define the norms of f and g in $\mathcal{H}_x$ and $\mathcal{H}_z$ as $||f||$ and $||g||$, and use these as regularization operators.

Next, we consider minimizing a regularized criterion defined as follows:

$$R(f,g) = \frac{\sum_{(\mathbf{x}_i,\mathbf{z}_j)\in E}(f(\mathbf{x}_i) - g(\mathbf{z}_j))^2 - \sum_{(\mathbf{x}_i,\mathbf{z}_j)\notin E}(f(\mathbf{x}_i) - g(\mathbf{z}_j))^2 + \lambda_1||f||^2 + \lambda_2||g||^2}{\sum_{(\mathbf{x}_i,\mathbf{z}_j)\in U\times V}(f(\mathbf{x}_i) - g(\mathbf{z}_j))^2}, \tag{3}$$

where λ_1 and λ_2 are regularization parameters that control the trade-off between minimization of the original objective function and smoothing of the functions.

To construct a multidimensional unified feature space of drugs and target proteins, we add orthogonality constraints and extract the m-th features f_m and g_m as $(f_m, g_m) = \arg\min R(f, g)$ recursively for $m = 1, \ldots, d$ under the orthogonality constraints: $f \perp f_1, \ldots, f_{m-1}$, and $g \perp g_1, \ldots, g_{m-1}$.

Applying the representer theorem in kernel methods [35], f_m and g_m can be written as linear combinations of kernel functions as follows: $f_m(\mathbf{x}) = \sum_{i=1}^{n_x} \alpha_{m,i} k_x(\mathbf{x}_i, \mathbf{x})$ and $g_m(\mathbf{z}) = \sum_{j=1}^{n_z} \beta_{m,j} k_z(\mathbf{z}_j, \mathbf{z})$ for some coefficients $\boldsymbol{\alpha}_m = (\alpha_{m,1}, \ldots, \alpha_{m,n_x})^T \in \mathbf{R}^{n_x}$ and $\boldsymbol{\beta}_m = (\beta_{m,1}, \ldots, \beta_{m,n_z})^T \in \mathbf{R}^{n_z}$. The optimization problem is then reduced to the generalized eigenvalue problem with respect to $\boldsymbol{\alpha}_m$ and $\boldsymbol{\beta}_m$, allowing f_m and g_m to be recovered. More details of this algorithm can be found in the original paper [27].

Finally, we evaluate the closeness between drugs and target proteins by calculating the inner product of their feature vectors in the unified feature space. Drug–protein pairs with high feature similarities are predicted to be candidates for interaction pairs. The method is referred to as kernel distance learning (KDL).

4 Results

4.1 Performance Evaluation of Metabolic Pathway Reconstruction

4.1.1 Data

A total of 13,564 enzymatic reactions were retrieved from the KEGG LIGAND database [36]. The compound–compound pairs in enzymatic reactions (substrate–product pairs) comprised the gold standard dataset.

The number of compounds for which structure information is available in KEGG LIGAND is 15,698, so that the number of possible compound–compound pairs is 246,411,506. The chemical structures of compounds were encoded using chemical fingerprints (binary vectors) whose elements represent the chemical substructures and physicochemical properties. Using the Chemistry Development Kit (CDK) [37], we derived five fingerprints: CDK, E-state, Klekota-Roth, MACCS, and PubChem. These had dimensions of 1024, 71, 4860, 164, and 881, respectively.

4.1.2 Experiments

We tested the ability of L1SVM to predict the enzymatic-reaction likeness and compared its performance with that of L2SVM. The following fivefold cross-validation was conducted. The compound–compound pairs in the gold standard data were split into five subsets of roughly equal size. Known reactant pairs were used as positive examples, and all other compound–compound pairs were used as negative examples. Each subset was used as the test set and the remaining four subsets were used as the training set. The predictive model was trained based only on the training set, and the prediction scores on the test set were calculated. The prediction accuracy was evaluated over the five tests.

The accuracy of the prediction was evaluated using the receiver operating characteristic (ROC) curve, which plots true positives as a function of false positives at various thresholds. If the area under the ROC curve (AUC) score returned 1, perfect inference was assumed, while 0.5 indicated random inference. The parameters were optimized using the AUC score as the objective function.

Table 1 shows the resulting AUC scores and their standard deviations. As a baseline (BASELINE), we used a standard similarity-based method, as the substrate–product pairs in known enzymatic reactions tend to have similar chemical structures [14]. L1SVM and L2SVM outperformed BASELINE on all fingerprints, suggesting that similarity in chemical structure does not necessarily imply enzymatic-reaction likeness. The PubChem fingerprint was found to be the most reliable. Although the L1SVM AUC scores were comparable to or slightly worse than those of L2SVM, L1SVM had a strong advantage over L2SVM in model interpretability. The total elements in the feature vectors when using the PubChem fingerprint numbered 2,643, and the numbers of features extracted (with nonzero weights) by L1SVM and L2SVM were around 650 and 1,700, respectively.

We examined highly weighted features corresponding to co-occurring (one formed and one eliminated) substructures. For example, in a highly weighted substructure labeled "C(~H) (~O)(~O)", a carbon atom attached with a hydrogen atom and two oxygen atoms, including a (hemi)acetal group, a (hemi)ketal group, a carboxyl group, an O-formyl group, etc., and was associated with the substructures labeled "C(~C)(~O)", including a hydroxy group, an aldehyde group, etc. The association between the two substructures is logical biochemically, as "aldehyde <=> hemiacetal", etc., transformations are common.

Table 1 AUC scores on fivefold cross-validation experiments for metabolic pathway reconstruction

Fingerprint	L1SVM	L2SVM	BASELINE	RANDOM
CDK	0.957 ± 0.001	0.942 ± 0.002	0.873 ± 0.004	0.500 ± 0.000
E-state	0.817 ± 0.005	0.777 ± 0.006	0.719 ± 0.008	0.500 ± 0.000
Klekota-Roth	0.951 ± 0.003	0.935 ± 0.004	0.854 ± 0.008	0.500 ± 0.000
MACCS	0.909 ± 0.002	0.902 ± 0.002	0.799 ± 0.007	0.500 ± 0.000
PubChem	0.952 ± 0.002	0.947 ± 0.003	0.871 ± 0.003	0.500 ± 0.000

4.2 Performance Evaluation of Drug–Target Interaction Prediction

4.2.1 Data

Drug–target interactions were obtained from the KEGG DRUG [38], SuperTarget [39], and DrugBank databases [40]. In this study, we focused on drug–target interactions in which the target proteins were enzymes. The number of drug–target interactions was 5,449, the number of target proteins was 1,062, and the number of drugs was 1,123. These data were used as the gold standard set.

We computed the kernel similarity value between the chemical structures of the drugs using the SIMCOMP algorithm [41], and the kernel similarity values of amino acid sequences between target proteins using the Smith–Waterman algorithms [42]. If a similarity score was not positive definite, an appropriate identity matrix was added to ensure that the corresponding kernel Gram matrix was positive definite, following a previous study [43].

4.2.2 Experiments

We tested the ability of KDL to predict drug–target interactions. As a baseline, we used the nearest neighbor method (NN), as this has been used in traditional approaches to molecular screening [44]. To compare the methods, we used kernel correspondence analysis (KCA), the kernelized version of correspondence analysis (CA) that is commonly used as an embedding method for heterogeneous objects in statistics [45].

We performed a fivefold cross-validation experiment, in which drugs and target proteins in the gold standard set were split into five subsets of roughly equal size. Each drug subset and target protein subset were used as test sets, and the remaining sets as training sets. The predictive model was trained only on the training sets, before being applied to the test sets. The performance was analyzed using the AUC score, calculated for different prediction classes depending on whether the drugs and/or the target proteins were present in the training set or not. Those in the training set were defined as "training drugs" and "training proteins", and those in the test sets as "test drugs" and "test proteins". Three prediction classes were used: (i) test drug versus training protein, (ii) training drug versus test protein, and (iii) test drug versus test proteins.

Table 2 shows the AUC scores and their standard deviations from the fivefold cross-validation experiments. KDL significantly outperformed KCA and NN. The inferior performance of NN suggests that similarities in raw drug structure and in protein sequence do not always reflect interactions. These results suggest that the feature space learned by KDL is able to capture the bipartite graph structure of the drug–target interaction network, more efficiently than that of KCA. Among the

Table 2 AUC scores on fivefold cross-validation experiments for drug–target interaction prediction

Drug target	Method	AUC		
		(i) test drugs versus training proteins	(ii) training drugs versus test proteins	(iii) test drugs versus test proteins
Enzyme	Random	0.500 ± 0.000	0.500 ± 0.000	0.500 ± 0.000
	NN	0.655 ± 0.011	0.758 ± 0.008	0.500 ± 0.000
	KCA	0.741 ± 0.011	0.839 ± 0.009	0.692 ± 0.008
	KDL	0.843 ± 0.006	0.878 ± 0.003	0.782 ± 0.013

prediction classes, the accuracy of the "test drug versus test protein" was weakest, because of the scarcity of knowledge on known interaction partners.

Finally, a comprehensive prediction was conducted of interactions between all drugs, including drug-like compounds, and all enzymes coded in the human genome. Many interesting cases emerged in drug–protein pairs with high prediction scores. For example, COX enzymes are a common target for anti-inflammatory drugs due to the role they play in the synthesis of prostanoids and the subsequent inflammation response [46]. 4-Hydroxyhydratropate and 2,2-Bis(4-hydroxyphenyl)-propanoic acid were predicted to interact with COX, neither of which, to our knowledge, have previously been identified as potential COX inhibitors. Another high prediction score was found for Imatinib mesylate, a tyrosine kinase inhibitor used in the treatment of chronic myelogenous leukemia and gastrointestinal tumors. Imatinib mesylate was predicted to interact with related tyrosine kinases including protein tyrosine kinase 6 (PTK6) and B-lymphoid tyrosine kinase, both of which are either confirmed or candidate oncogenes.

5 Discussion and Conclusions

In this study, we investigated recent advances in statistical machine learning methods for metabolic pathway reconstruction and drug–target interaction prediction, using heterogeneous biomedical big data. Recent omics studies have suggested the use of data-driven approaches for exploring previously unknown biosynthetic mechanisms in the metabolic pathways and molecular mechanisms of human disease. Statistical machine learning methods are expected to play an increasing role in rational compound synthesis and efficient drug discovery, and to increase the success rate of research in the food and pharmaceutical industries.

For metabolic pathway reconstruction, we introduced an L_1-regularized pairwise support vector machine to predict unknown enzymatic reactions among metabolome-scale compounds based on their compound structures, and demonstrated the usefulness of the method in terms of prediction accuracy and interpretability. The method will contribute to our understanding of the roles of metabolites and enzymes in the biosynthetic machinery. Computational prediction of unknown metabolic pathways is also expected to support experimental characterization in agricultural applications.

For prediction of drug–target interactions, we introduced supervised bipartite graph inference with kernel methods. These allowed previously unknown interactions between drugs and target proteins to be predicted, based on chemical and genomic data. The prediction accuracy and large-scale applicability of the method were confirmed. A comprehensive drug–target interaction network can identify many potential drug–target interactions. Fast computational screening of drug candidate compounds against therapeutic target proteins can dramatically reduce the experimental load and speed up the drug development process. The prediction of additional therapeutic target proteins for existing drugs will lead to the repositioning of existing drugs, marketing them applicable to a wide range of diseases [47].

Acknowledgements This work is supported by JST PRESTO Grant Number JPMJPR15D8, JSPS KAKENHI Grant Numbers 25700029 and 15K14980, and the Program to Disseminate Tenure Tracking System, MEXT, Japan and Kyushu University Interdisciplinary Programs in Education and Projects in Research Development.

References

1. Y. Toya, H. Shimizu, Flux analysis and metabolomics for systematic metabolic engineering of microorganisms. Biotechnol. Adv. **31**, 818–826 (2013)
2. D. Newman, G. Cragg, Natural products as sources of new drugs over the 30 years from 1981 to 2010. J. Nat. Prod. **75**, 311–335 (2012)
3. R. Nakabayashi, K. Saito, Metabolomics for unknown plant metabolites. Anal. Bioanal. Chem. **405**, 5005–5011 (2013)
4. F. Afendi, T. Okada, M. Yamazaki, A. Hirai-Morita, Y. Nakamura, K. Nakamura, S. Ikeda, H. Takahashi, M. Altaf-Ul-Amin, L. Darusman, K. Saito, S. Kanaya, KNApSAcK family databases: integrated metaboliteplant species databases for multifaceted plant research. Plant Cell Physiol. **53**, e1 (2012)
5. A. Sreekumar, L. Poisson, T. Rajendiran, A. Khan, Q. Cao, J. Yu, B. Laxman, R. Mehra, R. Lonigro, Y. Li, M. Nyati, A. Ahsan, S. Kalyana-Sundaram, B. Han, X. Cao, J. Byun, G. Omenn, D. Ghosh, S. Pennathur, D. Alexander, A. Berger, J. Shuster, J. Wei, S. Varambally, C. Beecher, A. Chinnaiyan, Metabolomic profiles delineate potential role for sarcosine in prostate cancer progression. Nature **457**, 910–914 (2009)
6. P. Karp, Call for an enzyme genomics initiative. Genome Biol. **5**, 401–401 (2004)
7. F. Darvas, Predicting metabolic pathways by logic programming. J. Mol. Graphics **6**, 80–86 (1988)
8. J. Talafous, L. Sayre, J. Mieyal, G. Klopman, A dictionary model of mammalian xenobiotic metabolism. J. Chem. Inf. Comput. Sci. **34**, 1326–1333 (1994)
9. N. Greene, P. Judson, J. Langowski, C. Marchant, Knowledge-based expert systems for toxicity and metabolism prediction: DEREK, StAR and METEOR. SAR QSAR Environ. Res. **10**, 299–314 (1999)
10. J. Faulon, A. Sault, Stochastic generator of chemical structure. 3. reaction network generation. J. Chem. Inf. Comput. Sci. **41**, 894–908 (2001)
11. L. Ellis, J. Gao, K. Fenner, L. Wackett, The University of Minnesota pathway prediction system: predicting metabolic logic. Nucleic Acids Res. **36**, W427–W432 (2008)

12. Y. Moriya, D. Shigemizu, M. Hattori, T. Tokimatsu, M. Kotera, S. Goto, M. Kanehisa, PathPred: an enzyme-catalyzed metabolic pathway prediction server. Nucleic Acids Res. **38**, W138–143 (2010)
13. V. Hatzimanikatis, C. Li, J. Ionita, C. Henry, M. Jankowski, L. Broadbelt, Exploring the diversity of complex metabolic networks. Bioinformatics **21**, 1603–1609 (2005)
14. M. Kotera, A. McDonald, S. Boyce, K. Tipton, Eliciting possible reaction equations and metabolic pathways involving orphan metabolites. J. Chem. Inf. Model. **48**, 2335–2349 (2008)
15. M. Nakamura, T. Hachiya, Y. Saito, K. Sato, Y. Sakakibara, An efficient algorithm for de novo predictions of biochemical pathways between chemical compounds. BMC Bioinform. **13** (2012)
16. M. Kotera, Y. Tabei, Y. Yamanishi, T. Tokimatsu, S. Goto, Supervised de novo reconstruction of metabolic pathways from metabolome-scale compound sets. Bioinformatics **29**, i135–i144 (2013)
17. M. Kotera, Y. Tabei, Y. Yamanishi, A. Muto, Y. Moriya, T. Tokimatsu, S. Goto, Metabolome-scale prediction of intermediate compounds in multistep metabolic pathways with a recursive supervised approach. Bioinformatics **30**, i165–i174 (2014)
18. Y. Yamanishi, Y. Tabei, M. Kotera, Metabolome-scale de novo pathway reconstruction using regioisomer-sensitive graph alignments. Bioinformatics **31**, i161–i170 (2015)
19. R. Ramautar, R. Berger, J. van der Greef, T. Hankemeier, Human metabolomics: strategies to understand biology. Cur. Opin. Chem. Biol. **17**, 841–846 (2013)
20. H. Lodhi, Y. Yamanishi, *Chemoinformatics and Advanced Machine Learning Perspectives: Complex Computational Methods and Collaborative Techniques* (IGI Global, 2010)
21. N. Nagamine, Y. Sakakibara, Statistical prediction of proteinchemical interactions based on chemical structure and mass spectrometry data. Bioinformatics **23**, 2004–2012 (2007)
22. J. Faulon, M. Misra, S. Martin, K. Sale, R. Sapra, Genome scale enzymemetabolite and drug-target interaction predictions using the signature molecular descriptor. Bioinformatics **24**, 225–233 (2008)
23. L. Jacob, J.-P. Vert, Protein-ligand interaction prediction: an improved chemogenomics approach. Bioinformatics **24**, 2149–2156 (2008)
24. K. Bleakley, Y. Yamanishi, Supervised prediction of drug-target interactions using bipartite local models. Bioinformatics **25**, 2397–2403 (2009)
25. Y. Tabei, E. Pauwels, V. Stoven, K. Takemoto, Y. Yamanishi, Identification of chemogenomic features from drug-target interaction networks using interpretable classifiers. Bioinformatics **28**, i487–i494 (2012)
26. Y. Yamanishi, M. Araki, A. Gutteridge, W. Honda, M. Kanehisa, Prediction of drug-target interaction networks from the integration of chemical and genomic spaces. Bioinformatics **24**, i232–i240 (2008)
27. Y. Yamanishi, Supervised bipartite graph inference. in *Advances in Neural Information Processing Systems 21*, ed. by D. Koller, D. Schuurmans, Y. Bengio, L. Bottou (MIT Press, Cambridge, MA, 2009), pp. 1841–1848
28. Y. Yamanishi, M. Kotera, M. Kanehisa, S. Goto, Drug-target interaction prediction from chemical, genomic and pharmacological data in an integrated framework. Bioinformatics **26**, i246–i254 (2010)
29. M. Takarabe, M. Kotera, Y. Nishimura, S. Goto, Y. Yamanishi, Drug target prediction using adverse event report systems: a pharmacogenomic approach. Bioinformatics **28**, i611–i618 (2012)
30. J. Zhu, 1-norm support vector machines, in *Advances in Neural Information Processing Systems 15*, ed. by S. Becker, S. Thrun, K. Obermayer (MIT Press, Cambridge, MA, 2003), pp. 49–56
31. R.E. Fan, K.W. Chang, C.J. Hsieh, X. Wang, C.J. Lin, LIBLINEAR: a library for large linear classification. J. Mach. Learn. Res. **9**, 1871–1874 (2008)
32. Y. Tabei, Y. Yamanishi, Scalable prediction of compound-protein interactions using minwise hashing. BMC Syst. Biol. **7**, S3 (2013)
33. G. Wahba, *Splines Models for Observational Data: Series in Applied Mathematics* (SIAM, Philadelphia, 1990)

34. F. Girosi, M. Jones, T. Poggio, Regularization theory and neural networks architectures. Neural Comput. **7**, 219–269 (1995)
35. J. Shawe-Taylor, N. Cristianini, *Kernel Methods for Pattern Analysis* (Cambridge University Press, London, 2004)
36. M. Kanehisa, S. Goto, Y. Sato, M. Furumichi, M. Tanabe, KEGG for integration and interpretation of large-scale molecular data sets. Nucleic Acids Res. **40**, D109–114 (2012)
37. C. Steinbeck, Y. Han, S. Kuhn, O. Horlacher, E. Luttmann, E. Willighagen, The chemistry development kit (CDK) an open-source Java library for Chemo- and bioinformatics. J. Chem. Inf. Comput. Sci. **43**, 493–500 (2003)
38. M. Kanehisa, S. Goto, M. Hattori, K. Aoki-Kinoshita, M. Itoh, S. Kawashima, T. Katayama, M. Araki, M. Hirakawa, From genomics to chemical genomics: new developments in kegg. Nucleic Acids Res. **34**, D354–357 (2006)
39. S. Gunther, S. Guenther, M. Kuhn, M. Dunkel et al., Supertarget and matador: resources for exploring drug-target relationships. Nucleic Acids Res **36**, D919–D922 (2008)
40. D. Wishart, C. Knox, A. Guo, D. Cheng, S. Shrivastava, D. Tzur, B. Gautam, M. Hassanali, Drugbank: a knowledgebase for drugs, drug actions and drug targets. Nucleic Acids Res **36**, D901–D906 (2008)
41. M. Hattori, Y. Okuno, S. Goto, M. Kanehisa, Development of a chemical structure comparison method for integrated analysis of chemical and genomic information in the metabolic pathways. J. Am. Chem. Soc. **125**, 11853–11865 (2003)
42. T. Smith, M. Waterman, Identification of common molecular subsequences. J. Mol. Biol. **147**, 195–197 (1981)
43. H. Saigo, J. Vert, N. Ueda, T. Akutsu, Protein homology detection using string alignment kernels. Bioinformatics **20**, 1682–1689 (2004)
44. D.T. Stanton, T.W. Morris, S. Roychoudhury, C.N. Parker, Application of nearest-neighbor and cluster analyses in pharmaceutical lead discovery. J. Chem. Inf. Comput. Sci. **39**, 21–27 (1999)
45. M. Greenacre, *Theory and Applications of Correspondence Analysis*. (Academic Press, 1984)
46. K. Rainsford, Anti-inflammatory drugs in the 21st century. Subcell. Biochem. **42**, 3–27 (2007)
47. R. Sawada, H. Iwata, S. Mizutani, Y. Yamanishi, Target-based drug repositioning using large-scale chemical-protein interactome data. J. Chem. Inf. Model. **55**, 27172730 (2015)

Compactly Supported Solutions of Reaction–Diffusion Models of Biological Spread

Maureen P. Edwards, Bronwyn H. Bradshaw-Hajek, María Jesús Munoz-Lopez, Peter M. Waterhouse and Robert S. Anderssen

Abstract Lie group analysis is one of the most useful techniques for analyzing the analytic structure of the solutions of differential equations. Here, reaction–diffusion (RD) modelling of biological invasion is used to illustrate this fact in terms of identifying the conditions that the diffusion and reaction terms must satisfy for their solutions to have compact support. Biological invasion, such as the spread of viruses on the leaves of plants and the invasive spread of animals and weeds into new environments, has a well-defined progressing compactly supported spatial $\mathbb{R}^2$ structure. There are two distinct ways in which such progressing compact structure can be modelled mathematically; namely, cellular automata modelling and reaction–diffusion (RD) equation modelling. The goal in this paper is to review the extensive literature on RD equations to investigate the extent to which RD equations are known to

---

M. P. Edwards
School of Mathematics and Applied Statistics, University of Wollongong,
Wollongong, NSW 2522, Australia
e-mail: maureen@uow.edu.au

B. H. Bradshaw-Hajek
School of Phenomics and Bioinformatics Research Centre, School of Information
Technology and Mathematical Sciences, University of South Australia,
Mawson Lakes Campus, Mawson Lakes, SA 5059, Australia
e-mail: Bronwyn.Hajek@unisa.edu.au

M. J. Munoz-Lopez
School of Mathematics, University of Minnesota, Minneapolis, MN 55455, USA
e-mail: munoz172@umn.edu

P. M. Waterhouse
Centre for Tropical Crops and Biocommodities, Queensland University of Technology,
Brisbane, QLD 4001, Australia
e-mail: peter.waterhouse@qut.edu.au

R. S. Anderssen (✉)
CSIRO Data61, GPO Box 664, Canberra, ACT 2601, Australia
e-mail: Bob.Anderssen@Data61.csiro.au

R. S. Anderssen
Department of Mathematics and Statistics, La Trobe University,
Melbourne, VIC 3086, Australia

R. S. Anderssen et al. (eds.), *Agriculture as a Metaphor for Creativity in All Human Endeavors*, Mathematics for Industry 28,
https://doi.org/10.1007/978-981-10-7811-8_13

have compactly supported solutions. Though the existence of compactly supported solutions of nonlinear diffusion equations, without reaction, is well documented, the conditions that the reaction terms should satisfy in conjunction with such nonlinear diffusion equations, for the compact support to be retained, has not been examined in specific detail. A possible partial connection relates to the results of Arrigo, Hill, Goard and Broadbridge, who examined, under various symmetry analysis assumptions, situations where the diffusion and reaction terms are connected by explicit relationships. However, it was not investigated whether the reaction terms generated by these relationships are such that the compact support of the solutions is maintained. Here, results from a computational analysis for the addition of different reaction terms to power law diffusion are presented and discussed. It appears that whether or not the reaction term is zero, as a function of its argument at zero, is an important consideration. In addition, it is confirmed algebraically and graphically that the shapes of compactly supported solutions are strongly controlled by the choice of the reaction term.

Keywords Biological invasion · Plant viruses · Gene silencing · Reaction–diffusion · Compactly supported solutions · Symmetry analysis

1 Introduction

In a biological invasion, the invader starts from the (finite) spatial location of the initial invasion/infection and progresses outwards with a finite velocity. It follows that the solution of any model formulated to simulate the invasion must be compactly supported, in that it is positive and finite on some compact spatial region, the boundary of which progresses outwards with a finite velocity, and is zero outside the boundary of this progressing spatial region. Here, the possibility is explored of using progressing compactly supported spatial solutions of reaction–diffusion (RD) equations

$$u_t = [k(u)u_x]_x + q(u), \tag{1}$$

where $k(u)$ and $q(u)$ denote the diffusion and reaction terms, respectively. The requirement that, for practical applications and to ensure realistic invasion models, the required solutions of (1) must be compactly supported has essentially been overlooked apart from a short acknowledgment in 2005 by Sander and Braddock [27] and more recently in Edwards et al. [10].

Specific early mention that nonlinear diffusion (only) equations (Eq. (1) with $q(u) = 0$) can have compactly supported solutions appears to date from King [18] in 1989, though, in 1959, Pattle [24], without any motivation, background or comment, published a general analytic solution for power law diffusion $k(u) = u^\sigma$, $\sigma > 0$, that contained compactly supported solutions as well as others.

The Pattle result established that some nonlinear diffusion equations can have compactly supported solutions, with King's paper [18] introducing new forms for $k(u)$.

For solutions used to model biological invasion, a crucial feature is that they must be spatially bounded. This type of solution is in contrast with a travelling wave solution, such as that for Fishers equation, since travelling wave solutions are positive except at infinity. This requirement can be satisfied by a set of solutions that is larger than the set that possesses compact support. For example, a solution that is positive in one portion of a region of interest, $u(x, t) > 0$ for $0 \leq x \leq x^*(t)$ and negative outside of this region, $u(x, t) < 0$ for $x > x^*(t)$ does not possess compact support. However, for practical application, one might 'stitch together' a solution by choosing $u(x, t)$ for $0 \leq x \leq x^*(t)$, and setting $u(x, t) = 0$ for $x > x^*(t)$. This procedure was implied but not discussed in [7]. A solution of this nature requires that $u(x, t) = 0$ is a solution of the corresponding RD equation, that, in turn, implies the requirement that $q(0) = 0$.

Consequently, $q(0) = 0$ is a necessary condition to guarantee compact support or compact support extensions. Computational simulations, for power law diffusion, indicated that $q(0) = 0$ is not always a sufficient condition and that, on occasions, some additional condition of the form $q(u) \geq k(u)$, in the neighbourhood of the origin, is required. In addition, for all the simulations tested when $q(0) \neq 0$, the compact support was not retained.

Here, the goal is the identification of conditions under which the addition of a reaction term $q(u)$ to such nonlinear diffusion (only) equations, such as those presented in Pattle [24] and King [18], retains compact support. Such information is important biologically, as the structure of the diffusion and the reaction terms encapsulate the interaction occurring between the spreading organism and the environment limiting the spatial spread. For example, for the spread of a virus on the leaf of a plant, the interaction is between the replicating virus and the immune gene silencing response of the plant predominantly at the boundary of the spread.

In part, the focus is a review of the earlier symmetry analysis literature about RD equations in terms of the choice of diffusion and reaction terms for which compactly supported solutions are known to exist.

We note here that an alternative to reaction diffusion equations for modelling biological invasions is the use of cellular automata. For example, cellular automata methods have been used by Mallet and Pillis [22] in the study of tumour-immune interactions, by Li and Yeh [21] in the modelling of changing land use and by Basse and Plank [4] in the modelling of weed invasion as a level set process. In these types of models, the compact support of the invasion is explicitly defined on a grid of cells with rules for how the invasion spreads from cell to cell. The challenge is the formulation of the rules that reflect the biology of the cellular spread. Though other choices have been used, the biologically appropriate choice of the grid is hexagonal as cells talk to their neighbours through boundaries. Topologically equivalent hexagonal patterns include brick walls.

The paper has been organized in the following manner. Background about biological invasion is given in Sect. 2 that includes an illustrative discussion of the spread of a virus on the leaf of a plant. The construction, using symmetry analysis, of compactly supported solutions of nonlinear diffusion equations is discussed in Sect. 3 with emphasis on power law diffusion. The corresponding situation for RD equations is the topic of Sect. 4 when the diffusion is power law. The existence of compactly supported solutions for general diffusion and reaction, in terms of the published symmetry analysis literature focusing on the publications of Broadbridge and colleagues [7, 13] and Arrigo and Hill [3], is examined in Sect. 5.

For some of the choices of $k(u)$ and $q(u)$, the structure of the compactly supported solutions is plotted graphically, in order to illustrate the considerable variability in the possible shapes that they can have. From a biological modelling perspective, it illustrates that the shapes that compactly supported solutions can have are strongly controlled by the choice of $q(u)$, highlighting the importance of the role being played by the reaction term in RD equations with nonlinear diffusion.

2 Biological Invasion and Spread

Because of its ecological, environmental and economic importance and consequences, biological invasion and spread are intensively researched topics [2, 11, 28]. Invasion occurs across all scales of biological activities from the spread of insects, such as the emerald ash borer [15], of weeds [4], of land use [21] and of plant viruses [10, 14]. In all these activities, the spread is compactly supported. Consequently, identifying conditions that guarantee that compactly supported solutions exist imposes quite strong limitations on how the modelling can be performed. In the sequel this is illustrated graphically in that, depending on the choices of $k(u)$ and $q(u)$, the structure of the compactly supported solutions can vary considerably.

The two key possibilities for modelling such situations are cellular automata (discussed briefly in the Introduction) and RD equation modelling (the topic of this paper).

2.1 *Viral Spread on the Leaf of a Plant*

A major goal in plant breeding is the production of new disease-resistant varieties. To this end, it is helpful to understand and model the dynamics of host–pathogen interactions and identify the genes involved.

Because of the impact plant viruses have on the productivity of food crops, they have been intensively studied [26]. An example of the spread of a virus on the leaf of a plant is given in Fig. 1a, that highlights the intensity of the activity occurring at the boundary of the viral spread. A similar spread on the surface of tomatoes is given in Fig. 1b.

Although the underlying mechanism of the plant's antiviral response to infection is reasonably well understood [1], the competition between the virus and the host's

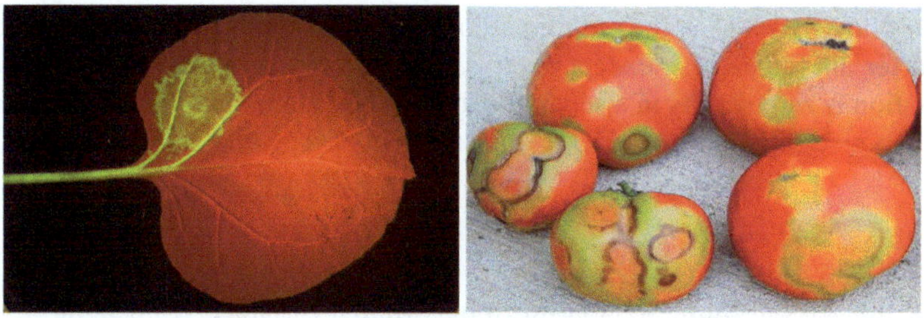

(a) (b)

Fig. 1 The spread of a fluorescence-tagged virus (Potato virus X), from the central inoculation spot, as an expanding circular-like front. When the front reaches major veins, the virus travels through them to exit the leaf

response at the boundary has not been modelled. We assume that the speed of a virus spreading from cell to cell is dictated by the balance between the replication rate of the virus and the plant's RNA interference mechanism inhibiting this replication.

3 Compactly Supported Solutions of Nonlinear Diffusion Equations

For the 1D nonlinear diffusion equation

$$u_t = [k(u)u_x]_x \tag{2}$$

compactly supported solutions have been derived for $k(u) = u^\sigma$ when $\sigma > 1$ [17, 18, 24, 29].

The symmetry analysis of Eq. (2) with $K(u) = u^\sigma$ ($\sigma \neq 4/3$) [17] shows that Eq. (2) admits symmetries characterized by the generators

$$\Gamma_1 = \frac{\partial}{\partial t}, \quad \Gamma_2 = \frac{\partial}{\partial x}, \quad \Gamma_3 = 2t\frac{\partial}{\partial t} + x\frac{\partial}{\partial x}, \quad \Gamma_4 = \frac{\sigma}{2}x\frac{\partial}{\partial x} + u\frac{\partial}{\partial u}.$$

A wider class of solutions can be derived by combining the two symmetries Γ_3 and Γ_4 to give

$$\Gamma = 2t\frac{\partial}{\partial t} + x\frac{\partial}{\partial x} + 2A\left(\frac{\sigma}{2}x\frac{\partial}{\partial x} + u\frac{\partial}{\partial u}\right), \quad A \in \mathbb{R}.$$

This yields the invariants

$$u = t^A F(\eta), \quad \eta = \frac{x}{t^{(A\sigma+1)/2}}.$$

For the choice $A = \frac{-1}{\sigma+2}$, the reduced second-order differential equation can be integrated for all values of σ except $\sigma = -2$. Assuming that the arbitrary constant of integration is zero results in a separable first-order differential equation that may be solved to give (as detailed in [10])

$$F^{\sigma} = \frac{A\sigma}{2}\eta^2 + K, \quad K \in \mathbb{R}.$$

Hence, the explicit solution, in terms of the original variables, is given by

$$u(x,t) = t^A F(\eta) = t^{\frac{-1}{(\sigma+2)}} \left[K - \frac{\sigma}{2(\sigma+2)} \frac{x^2}{t^{2/(\sigma+2)}} \right]^{1/\sigma}. \tag{3}$$

We note that $u(x^*, t) = 0$ when

$$x^* = t^{1/(\sigma+2)} \sqrt{\frac{2K(\sigma+2)}{\sigma}}. \tag{4}$$

Hence, as t increases, the value of x^* increases, ensuring the compactly supported region expands.

The solution (3) is illustrated in Fig. 2 with $K = 5$, and (a) $\sigma = 1$ for varying t, and (b) $t = 0.5$ for varying σ. The solution illustrated in Fig. 2a does not have compact support, as we have chosen $\sigma = 1$ in plot (a) of Fig. 2. Consequently, there is no restriction that

$$K - \frac{\sigma}{2(\sigma+2)} \frac{x^2}{t^{2/(\sigma+2)}}$$

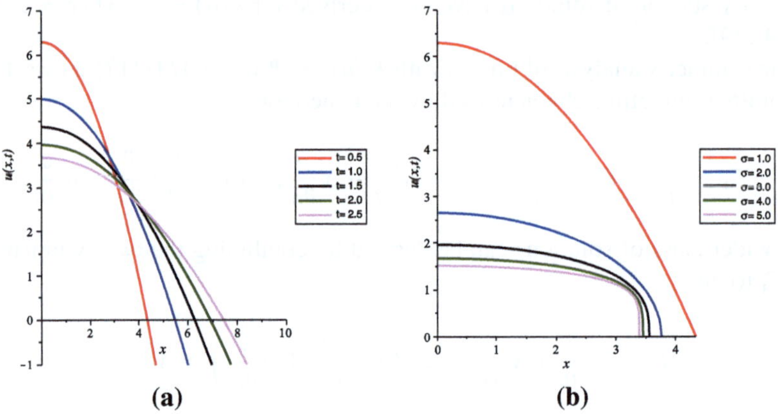

Fig. 2 **a** Illustration of Pattle solution (3) for $\sigma = 1$ and varying t. **b** Illustration of Pattle solution (3) for $t = 0.5$

(i.e. the quantity in the square brackets in Eq. (3)) cannot be negative, as shown in Fig. 2a. The compact solution generated, when we choose solution (3) for $x \leq x^*(t)$ and $u(x,t) = 0$ for $x > x^*(t)$, may still be biologically relevant.

Choosing $\sigma \neq 2n+1$, $(n \in \mathbb{N})$ will ensure that solution (3) is real and positive for $x \leq x^*$, guaranteeing a compactly supported structure. From Fig. 2b, it is clear that the solution approaches an asymptote as σ increases. Indirect confirmation includes the fact that

$$\lim_{\sigma\to\infty} u(0,t) = 1,$$

and, from (4), that $\lim_{\sigma\to\infty} x^* = \sqrt{2K}$.

Solution (3) is equivalent to the Pattle solution [24], and has also been examined by authors such as Hill [16] and Ibragimov [17]. Philip and Knight [25] generated the Pattle solution using a similarity solution argument.

4 Compactly Supported Solutions of Nonlinear Reaction–Diffusion Equations

The nonlinear diffusion equation (2) is now extended to include either a linear or a nonlinear reaction term $q(u)$ to give Eq. (1); namely,

$$u_t = [k(u)u_x]_x + q(u).$$

Given we know that compactly supported solutions of (2) with $k(u) = u^\sigma$ exist, what is the consequence of adding a *nonlinear* reaction term?

We already know that $q(0) = 0$ is a necessary condition for compactly supported solutions to exist, and it appears to be a sufficient condition in some circumstances. Here, classical symmetry analysis is used to explore the consequences of adding various reaction terms. Results for general diffusion and reaction are discussed in Sect. 5.

The Lie symmetry group method [20] is used to obtain the Lie point symmetry generators of differential equations. A one-parameter group of transformations

$$\begin{aligned} x^* &= x + \epsilon X(x,t,u) + O(\epsilon^2) \\ t^* &= t + \epsilon T(x,t,u) + O(\epsilon^2) \\ u^* &= u + \epsilon U(x,t,u) + O(\epsilon^2) \end{aligned} \tag{5}$$

that leaves the 1+1 dimensional partial differential equation (1) invariant is sought. The second prolongation of the group (5) is

$$\Gamma^{(2)} = X\frac{\partial}{\partial x} + T\frac{\partial}{\partial t} + U\frac{\partial}{\partial u} + U_{[x]}\frac{\partial}{\partial u_x} + U_{[t]}\frac{\partial}{\partial u_t} + U_{[xx]}\frac{\partial}{\partial u_{xx}} + +U_{[xt]}\frac{\partial}{\partial u_{xt}} + +U_{[tt]}\frac{\partial}{\partial u_{tt}} \tag{6}$$

Table 1 Forms of $k(u)$ and $q(u)$ that admit additional symmetries, $|\alpha| = |\delta| = 1$

$k(u)$	$q(u)$
e^u	$\pm e^{\beta u}, \ \pm e^u + \delta$
$u^\sigma, \sigma \neq 0, -4/3$	$\pm u^n, \ \pm u^{\sigma+1} + \delta u$
$u^{-4/3}$	$\pm u^n \, (n \neq -1/3), \ \alpha u^{-1/3} + \delta u \, (5 \; symmetries), \ \alpha u^{-1/3} \, (5 \; symmetries)$
1	$\pm e^u, \ \pm u^n, \ \delta u \ln u \, (4 \; symmetries)$

where, for example, $U_{[x]} = D_x(U) - D_x(X)u_x - D_x(T)u_t$, with D_x the total derivative

$$D_x = \frac{\partial}{\partial x} + u_x \frac{\partial}{\partial u} + u_{xx} \frac{\partial}{\partial u_x} + u_{xt} \frac{\partial}{\partial u_t} + u_{xxx} \frac{\partial}{\partial u_{xx}} + \ldots .$$

The references [5, 23] are recommended for a more complete description of Lie symmetry techniques. Applying the second prolongation of the symmetry generator (6) to the reaction diffusion equation (1), and then using the governing equation (1) to eliminate the derivative u_{xx} leads to a set of overdetermined partial differential equations (the determining equations) for the infinitesimals $X(x, t, u)$, $T(x, t, u)$ and $U(x, t, u)$. The classical symmetry classification of (1) is listed in Ibragimov [17]. Only translations in x and t are admitted for arbitrary forms of $k(u)$ and $q(u)$. Additional symmetries are admitted only when $k(u) = u^\sigma$ or $k(u) = e^u$.

Table 1 lists the various forms of $k(u)$ and associated $q(u)$ that admit additional symmetries. Each of the cases tabulated has a total of three symmetries unless otherwise indicated. Some forms of the RD equations which can be transformed into simplified RD equations have been omitted from Table 1. More information regarding the specific cases can be found in Ibragimov [17].

In the previous section, the symmetry properties of the diffusion equation (2) were exploited to generate compactly supported solutions. The technique involved taking a linear combination of symmetries to find invariants, generating the functional form and reducing the governing equation to an ordinary differential equation. Solving this reduced equation led to the explicit solution (3). A compactly supported solution is possible using the classical symmetry properties of the RD equation when $k(u) = u^\sigma$ and $q(u) = u^{\sigma+1}$, with $\sigma > 0$ [17]. In this solution, the position of the front is independent of t, and so does not have an expanding boundary.

5 General Diffusion and Reaction

Similar results to those discussed above hold for the existence and shapes of compact solutions generated by a nonclassical symmetry analysis of the types of RD equations examined here. For nonclassical symmetries, we look for a group of transformations, such as those described above, that leave Eq. (1) invariant, and that also satisfy the invariant surface condition [6],

$$u_t T + u_x X = U.$$

The earlier work of Arrigo and Hill [3] can be used to assist with the identification of conditions on the diffusion and reaction terms that guarantee the existence of compactly supported solutions.

Here, we consider the more general RD situation as, within this framework, partial answers can be identified to how the compactly supported solutions of nonlinear diffusion equations, such as the power law, are affected on the addition of reaction terms.

Arrigo and Hill [3] and Goard and Broadbridge [13] list a nonclassical symmetry property of (1) with the reaction term $q(u)$ related to a general diffusion term $k(u)$. The Arrigo and Hill analysis shows that, on assuming that

$$T(x,t,u) = 1, \quad X(x,t,u) = 0, \quad U(x,t,u) = U(u),$$

the following nonclassical symmetry is obtained

$$\Gamma = \frac{\partial}{\partial t} + \frac{c_1 \int k(u)\,du + c_2}{k(u)} \frac{\partial}{\partial u}$$

with the constraint

$$q(u) = (1 + c_3 k(u)) \left[\frac{c_1 \int k(u)\,du + c_2}{k(u)} \right]. \tag{7}$$

The corresponding characteristic equations are

$$\frac{dt}{1} = \frac{dx}{0} = \frac{k(u)\,du}{c_1 \int k(u)\,du + c_2}$$

that lead to the invariants

$$\eta = x, \qquad F(u) = \int k(u)\,du = \begin{cases} f(\eta)e^{c_1 t} - \frac{c_2}{c_1}, & c_1 \neq 0, \\ f(\eta) + c_2 t, & c_1 = 0. \end{cases}$$

The reaction–diffusion (1) equation reduces to

$$f'' + c_1 c_3 f = 0 \;\; (c_1 \neq 0) \qquad \text{or} \qquad f'' + c_2 c_3 = 0 \;\; (c_1 = 0).$$

Exact solutions for arbitrary $k(u)$ are [3]

$$u(x,t) = \begin{cases} F^{-1}\left(A\sin(\omega(x-x_0))\exp(c_1 t) - c_2/c_1\right), & c_1 \neq 0, \ \omega^2 = c_1 c_3 > 0, \\ F^{-1}\left(A(x-x_0)\exp(c_1 t) - c_2/c_1\right), & c_1 \neq 0, \ \omega^2 = c_1 c_3 = 0, \\ F^{-1}\left(A\sinh(\bar{\omega}(x-x_0))\exp(c_1 t) - c_2/c_1\right), & c_1 \neq 0, \ \bar{\omega}^2 = -c_1 c_3 > 0, \\ F^{-1}\left(-\frac{1}{2}c_2 c_3 x^2 + A(x-x_0) - c_2 t\right), & c_1 = 0. \end{cases} \tag{8}$$

Choosing $k(u) = e^u$ or $k(u) = u^m$ and using (7) to find the corresponding form of $q(u)$ will recover the cases already presented by Arrigo and Hill [3]. Other choices of $k(u)$ with the appropriate form of $q(u)$ will lead to other solutions to (1). For example, Arrigo and Hill [3] generate solutions for diffusion $k(u) = (1 - u^2)^{-1}$, with corresponding reaction term $q(u) = (c_3 + 1 - u^2)\left(c_1 \tanh^{-1}(u) + c_2\right)$.

There are two approaches to exploiting the constraint (7) to identify potentially compactly supported solutions of the RD equation (1). We could choose the form of diffusion $k(u)$ and use (7) to find the corresponding reaction term $q(u)$ and then test to see if the condition $q(0) = 0$ is satisfied. Alternatively, we could choose the reaction term $q(u)$ such that $q(0) = 0$, and use (7) to find the corresponding diffusion term. We investigate the second option. Let

$$F(u) = \int k(u)\, du \quad \left(\text{that is, } \frac{dF}{du} = k(u)\right),$$

and rearrange (7) to give the first-order differential equation

$$\frac{dF}{du} = \frac{c_1 F(u) + c_2}{q(u) - c_3(c_1 F(u) + c_2)}. \tag{9}$$

For a given reaction term $q(u)$, solving (9) will give the appropriate form of $k(u)$. If $c_3 = 0$, the differential equation (9) is separable (linear homogenous) and can be solved to give

$$F(u) = \frac{1}{c_1}\left[a_1 \exp\left(\int \frac{c_1}{q(u)}\, du\right) - c_2\right], \quad c_1 \neq 0,$$

with

$$k(u) = \frac{dF}{du} = \frac{a_1}{q(u)} \exp\left(\int \frac{c_1}{q(u)}\, du\right).$$

From [3], we have

$$F(u) = -\frac{c_2}{c_1} + f(x)\exp(c_1 t)$$

with $f'' = 0$ (since $c_3 = 0$), and so $f(x) = Ax + b = A(x - x_0)$. In general, when $c_3 = 0$,

$$\int \frac{c_1}{q(u)}\, du = \ln\left[\frac{A}{a_1}(x - x_0)e^{c_1 t}\right], \quad c_1, a_1 \neq 0.$$

Example A: $q(u) = u(1-u)$

In this case,

$$F(u) = \frac{a_1}{c_1}\left(\frac{u}{u-1}\right)^{c_1} - \frac{c_2}{c_1}, \qquad k(u) = -a_1\frac{u^{c_1-1}}{(u-1)^{c_1+1}}$$

and

$$u(x,t) = \frac{\left(\frac{c_1 A}{a_1}(x-x_0)\right)^{1/c_1} e^t}{\left(\frac{c_1 A}{a_1}(x-x_0)\right)^{1/c_1} e^t - 1}, \quad c_1, a_1 \neq 0. \tag{10}$$

For such a $u(x,t)$ to be compactly supported, conditions are required on c_1, A and a_1 such that $u(x,t) > 0$ for $0 < x < x_0$ and $u(x,t) = 0$ when $x \geq x_0$. When $c_1 = 1$,

$$u(x,t) = \frac{\frac{A}{a_1}(x-x_0)e^t}{\frac{A}{a_1}(x-x_0)e^t - 1}, \quad a_1 \neq 0,$$

and the solution will always have a fixed stationary boundary (i.e. $u = 0$ at $x = x_0$). Since $x - x_0 < 0$ in the region of interest, choosing $a_1 A > 0$ will give $u(x,t) > 0$ for $x < x_0$. The progressive dynamics of this type of solution is illustrated in Fig. 3a.

When $c_1 > 1$, we need

$$\frac{c_1 A}{a_1}(x - x_0) > 0.$$

Since $x - x_0 < 0$, we need $a_1 A < 0$. However, because of the form of the denominator in equation (10), $u(x,t)$ will always have a singularity x_s such that $0 < x_s < x_0$.

When $0 < c_1 < 1$, choosing $\frac{A}{a_1} > 0$ will give a solution when, for example, $c_1 = \frac{1}{2n+1}$, since the denominator in (10) will always be negative. When $c_1 = \frac{1}{2n}$, the denominator will, depending on the value of $\frac{A}{a_1} > 0$, change sign indicating that the solution will have a singularity in the range $0 < x < x_0$. If $1/c_1$ is not an integer, then the solution is complex for $0 < x < x_0$.

An example of the progressive nature of this solution is illustrated in Fig. 3b with $c_1 = \frac{1}{3}$. When $x > x_0$, such solutions become negative before reaching a singularity (not shown in figure).

Finally, when $c_1 < 0$, there is no x such that $u(x,t) = 0$, so no compactly supported solution exists in this case. The solutions shown in Fig. 3 are not compactly supported, since they are defined for $x > x_0$. However, their compact forms may be of biological relevance since they have well-defined boundaries.

The considerable variability in the structure of such solutions again yields validation of the important role played by the reaction term in RD equations.

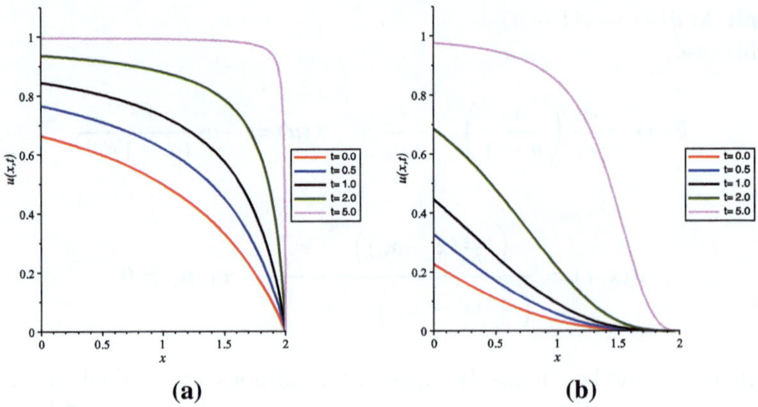

Fig. 3 Illustration of solution (10) with $x_0 = 2$, $A = a_1 = 1$ and **a** $c_1 = 1$, **b** $c_1 = \frac{1}{3}$

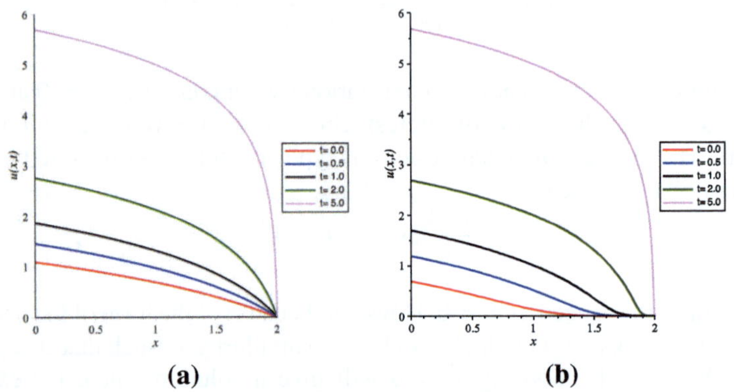

Fig. 4 Illustration of solution (11) with $x_0 = 2$, $A = -1$, $a_1 = 1$ and **a** $c_1 = \alpha = 1$, **b** $c_1 = 1$ and $\alpha = 5$

Example B: $q(u) = 1 - e^{-\alpha u}, \alpha > 0$

In this case,

$$u(x,t) = \frac{1}{\alpha} \ln \left[\left(\frac{A}{a_1}(x - x_0) \right)^{\alpha/c_1} e^{\alpha t} + 1 \right] \tag{11}$$

and

$$k(u) = a_1 e^{\alpha u} \left(e^{\alpha u} - 1 \right)^{(c_1 - 1)/\alpha}.$$

We can satisfy $u(x,t) = 0$ when $x = x_0$ and $c_1 > 0$ (as $\alpha > 0$), and $u(x,t) > 0$ in the region $x < x_0$ when $a_1 A < 0$. The solution is illustrated for different combinations of α, c_1, A and a_1 in Fig. 4.

6 Final Remarks

Here, reaction–diffusion (RD) modelling of biological invasion has been used to illustrate the usefulness of Lie group analysis in the recovery of information about the analytic structure of the solutions of differential equations. Here, the focus was an exploration of the RD equation literature to review and identify conditions under which RD equations have compactly supported solutions, since only such solutions are appropriate models of biological invasions.

The importance and relevance of having $q(0) = 0$, a necessary condition for the existence of solutions with compact support, is discussed from a number of perspectives including the results from computational simulations.

It was confirmed algebraically and graphically that the shapes of compactly supported RD solutions are strongly controlled by the choice of the reaction term with a considerable variability in structure with some having very definite precipitate boundaries while others have slowly smoothly decaying boundaries.

The possibility for future research is highlighted in Sect. 5, with particular mention of the relationship (7), in order to derive conditions on $q(u)$ such that compactly supported solutions of nonlinear diffusion equations are retained when the diffusion equation is transformed to be a reaction diffusion equation through the addition of the $q(u)$.

Acknowledgements The authors would like to acknowledge the discussion with Rick Loy that led to introducing the condition that $q(0)$ must equal zero in order to guarantee compactly supported solutions. RSA, BH and ME greatly appreciate the financial support of the Institute for Mathematics and Its Applications (IMIA) at the University of Wollongong which has underpinned their collaboration related to the research reported here. They also acknowledge the important mentoring that they have received over the years from Phil Broadbridge.

References

1. R.S. Anderssen, P.M. Waterhouse, Antiviral resistance in plants in *Modelling Antiviral Resistance in Plants - Methods in Molecular 894*, ed. by J. Watson, M.-B. Wang (Humana Press, 2012), pp. 139–140
2. M. Arim, S.R. Abades, P.E. Neill, M. Lima, P.A. Marquet, Spread dynamics of invasive species. PNAS **103**, 374–378 (2006)
3. D.J. Arrigo, J.M. Hill, Nonclassical symmetries for nonlinear diffusion and absorption. Stud. Appl. Math. **94**, 21–39 (1995)
4. B. Basse, M. Plank, Modelling biological invasions over homogeneous and inhomogeneous landscapes using level set methods. Biol. Invasions **10**, 157–167 (2008)
5. G.W. Bluman, S. Kumei, *Symmetries and Differential Equations* (Springer, Berlin, 1989)
6. G.W. Bluman, J.D. Cole, General similarity solution of the heat equation. J. Math. Mech. **18**, 1025–1042 (1996)
7. P. Broadbridge, B.H. Bradshaw-Hajek, Exact solutions for logistic reaction-diffusion in biology. Zeitschrift Angew. Mathematik Phys. **67**, Article 93 (2016)
8. B.L. Cheeseman, D.F. Newgreen, K.A. Landman, Spatial and temporal dynamics of cell generations within an invasion wave: a link to cell lineage tracing. J. Theor. Biol. **363**, 344–356 (2014)

9. A. Eamens, M.-B. Wang, N.A. Smith, P.M. Waterhouse, RNA silencing in plants: yesterday, today, and tomorrow. Plant Physiol. **147**, 456–468 (2008)
10. M.P. Edwards, P.M. Waterhouse, M.J. Munoz-Lopez, R.S. Anderssen, Nonlinear diffusion and viral spread through the leaf of a plant. Zeitschrift Angew. Mathematik Phys. **67**, Article 112 (2016)
11. W.F. Fagan, M.A. Lewis, M.G. Neubert, P. Van Den Driessche, Invasion theory and biological control. Ecol. Lett. **5**, 148–157 (2002)
12. R.A. Fisher, The wave of advance of advantageous genes. Ann. Eugen. **7**, 355–369 (1937)
13. J. Goard, P. Broadbridge, Nonclassical symmetry analysis of nonlinear reaction-diffusion equations in two spatial dimensions. Nonlinear Anal. Theory Methods Appl. **26**, 735–754 (1996)
14. M.A.C. Groenenboom, P. Hogeweg, RNA silencing can explain chlorotic infection patterns on plant leaves. BMC Syst. Biol. **2**, Article 105 (2008)
15. D.A. Herms, D.G. McCullough, Emerald ash borer invasion of North America: history, biology, ecology, impacts, and management. Annu. Rev. Entomol. **59**, 13–30 (2014)
16. J.M. Hill, *Differential Equations and Group Methods for Scientists and Engineers* (CRC Press, Boca Raton, 1992)
17. N.H. Ibragimov, *CRC Handbook of Lie Group Analysis of Differential Equations, Volume I: Symmetries, Exact Solutions, and Conservation Laws* (CRC Press, Boca Raton, 1994)
18. J.R. King, Exact solutions to some nonlinear diffusion equations. Q. J. Mech. Appl. Math. **42**, 407–409 (1989)
19. J.C. Larkin, N. Young, M. Prigge, M.D. Marks, The control of trichome spacing and number in Arabidopsis. Development **122**, 997–1005 (1996)
20. S. Lie, Uber die Integration durch bestimmte Integrale von einer Klasse linearer partieller Differentialgleichungen. Arch. Math. **6**, 328–368 (1881)
21. X. Li, A.G.-O. Yeh, Neural-network-based cellular automata for simulating multiple land use changes using GIS. Int. J. Geogr. Inform. Sci. **16**, 323–343 (2002)
22. D.G. Mallet, L.G. De Pillis, A cellular automata model of tumor-immune system interactions. J. Theor. Biol. **239**, 334–350 (2006)
23. P.J. Olver, *Applications of Lie Groups to differential equations* (Springer, New York, 1993)
24. R.E. Pattle, Diffusion from an instantaneous point source with a concentration-dependent coefficient. Q. J. Mech. Appl. Math. **12**, 407–409 (1959)
25. J.R. Philip, J.H. Knight, Redistribution of soil water from plane, line, and point sources. Irrig. Sci. **12**, 169–180 (1991)
26. E.P. Rybicki, A Top Ten list for economically important plant viruses. Arch. Virol. **160**, 17–20 (2015)
27. G.C. Sander, R.D. Braddock, Analytical solutions to the transient, unsaturated transport of water and contaminants through horizontal porous media. Adv. Water Resour. **28**, 1102–1111 (2005)
28. N. Shigesada, K. Kawaasaki, *Biological Invasion: Theory and Practice* (Oxford University Press, 1997)
29. Ya.B. Zeldovich, A.S. Kompaneets, On the theory of propagation of heat with the heat conductivity depending upon the temperature, in *Collection in Honor of the Seventieth Birthday of Academician*, ed. by A.F. Ioffe (1950), pp. 61–71

Two Challenging Difficulties of Protein Complex Prediction

Osamu Maruyama

Abstract A protein complex is a group of proteins which carries out particular functions in the cell. The component proteins of a protein complex are connected via weak physical contacts, called protein–protein interactions (PPIs). Proteome-wide PPIs are determined by high-throughput assays. Thus, it is interesting to computationally predict protein complexes from such PPIs. In this paper, we describe two challenging difficulties of the problem. The first difficulty is that the smallest protein complexes are of size two. It is quite difficult to predict them due to their simple inherent structure. The second difficulty is that some known complexes are overlapped with each other, because it is not trivial how to model such overlaps mathematically. For these issues, we have proposed our own approaches. In both methods, we design a scoring function and algorithms based on Markov chain Monte Carlo to optimize the scoring function. In this article, we briefly show our key regularization terms included in the whole scoring function.

Keywords Protein complex · Protein–protein interaction · Markov chain Monte Carlo

1 Protein Complex Prediction

Every protein is described to have its own biological function. However, the function of a protein is executed when it forms an appropriate protein complex with other proteins. Thus, protein complexes are the mandatory entities in the cell to describe various biological process and cellular mechanisms. However, it is tedious and time-consuming to determine all protein complexes of an organism by biological assays. Unfortunately, at this moment, we have no reliable high-throughput assays to detect protein complexes.

O. Maruyama (✉)
Institute of Mathematics for Industry, Kyushu University, Fukuoka, Japan
e-mail: om@imi.kyushu-u.ac.jp

R. S. Anderssen et al. (eds.), *Agriculture as a Metaphor for Creativity in All Human Endeavors*, Mathematics for Industry 28,
https://doi.org/10.1007/978-981-10-7811-8_14

Thus, it is interesting to *computationally* predict protein complexes from protein–protein interactions (PPIs). PPIs are weak physical contacts between proteins, by which a protein complex is formed. With the advent of accumulated PPIs, determined by high-throughput assays, especially the yeast two-hybrid (Y2H) assay, we can predict protein complexes from those PPIs.

However, PPIs are noisy as most of them are determined by high-throughput assays. This makes protein complex prediction more difficult. In addition to the difficulty, we have two more difficulties. One is that the smallest protein complex is of size two. Trivially, it is relatively difficult to detect them due to their simple inherent structure. The other is that some known protein complexes are overlapped with each other, i.e., share the same one or more proteins. A mathematical model for overlapping complexes can bring us more precise detection of them.

In this article, we describe our two methods, which deal with the above two issues. Our method, PPSampler2 [1], deals with the first difficulty by using a regularization term to control the distribution of sizes of predicted complexes. For the second difficulty, we have proposed a new method, RocSampler [2], which allows predicted complexes to overlap with each other. This article explains those difficulties and briefly show our key idea for them.

2 Preliminaries

A PPI network is represented by an edge-weighted undirected graph, $G = (V, E, w)$, where V is a set of proteins, E is a set of interactions between proteins, and $w : V^2 \to R_{\geq 0}$ is a mapping from a pair of proteins to a nonnegative real, representing a weight of a PPI. For a pair of proteins, $\{u, v\}$, not included in E, $w(u, v)$ is set to be 0. In our work, a PPI database of yeast, WI-PHI [3], is used, in which the weights of the PPIs are called "reliability." Those data are derived from various heterogeneous data sources, including results of tandem affinity purification coupled to mass spectrometry (TAP-MS), large-scale yeast two-hybrid studies, and small-scale experiments stored in dedicated databases.

For a predicted complex, x, we denote by $w(x)$ the sum of the weights of all internal PPIs within x, i.e., $w(x) = \sum_{u,v \in x} w(u, v)$. A partition of all proteins of G is a set of subsets, $x_1, x_2, \ldots, x_m$, of V such that $x_i \cap x_j = \emptyset$ for any i and j, and $\cup_{i=1}^{m} x_i = V$.

3 Small Protein Complexes

The latest and most popular database of protein complexes of yeast is CYC2008 [4]. This database includes 408 protein complexes, among which 172 (42%) complexes are of size two and 87 (21%) complexes are of size three. Thus, these small complexes are more than half of all the complexes.

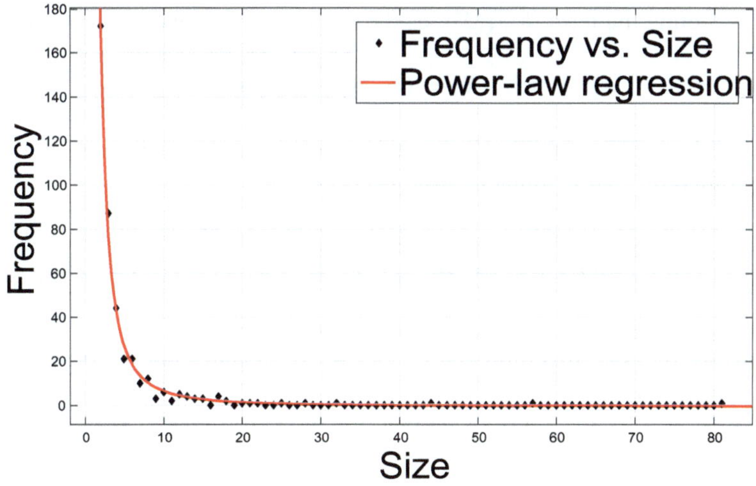

Fig. 1 A dot represents the frequency of a particular size of protein complexes. The curve is a power-law regression curve

Trivially, it is much harder to predict those small complexes than larger ones due to their simpler internal PPI structure. However, we have the helpful prior knowledge that the sizes of the known complexes of the database are distributed according to a power-law. The sizes of the complexes of CYC2008 is ranged from 2 to 81. The power-law regression curve for the data set, shown in Fig. 1, is proportional to $s^{-2.02}$ with $s \in [2, 100]$, with root-mean-square error 1.75. Thus, the scaling exponent is 2.02. Note that scaling exponents of power-law distributions in nature are typically in the range from two to three [5].

Our idea is to design a prior distribution of sizes of predicted complexes based on this observation. We have formulated a regularization term to fit the distribution of sizes of predicted complexes to a power-law distribution. Our prediction method, PPSampler2 [1], generates a partition, X, of V. All elements of size two or more of X are considered to be predicted complexes. The probability function for cluster size, s, of the two-sided truncated power-law distribution with a scaling exponent, γ, over the range $[2, S_{\max}]$ is

$$\psi_{\gamma}(s) = \frac{s^{-\gamma}}{\sum_{t=2}^{S_{\max}} t^{-\gamma}},$$

where $S_{\max}$ is the upper bound on the size of predicted complexes, and $s = 2, 3, \ldots, S_{\max}$. The default value of $S_{\max}$ is simply 100. We denote by $\psi_X(s)$ the fraction of predicted clusters of size s in X, i.e.,

$$\psi_X(s) = \frac{|\{x \in X | |x| = s\}|}{|\{x \in X | |x| \geq 2\}|}.$$

Then, we define the term, $h_{cs,s}(X)$, as the square error between $\psi_X(s)$ and $\psi_\gamma(s)$, i.e.,

$$h_{cs,s}(X) = (\psi_X(s) - \psi_\gamma(s))^2.$$

Our method, PPSampler2 [1], uses this regularization term as part of the whole scoring function, $f(X)$, which is formulated as

$$f(X) = b(X) + h_d(X) + c_{cs} \cdot \sum_{s=2}^{S_{\max}} h_{cs,s}(X, \gamma) + c_{pn} \cdot h_{pn}(X).$$

Note that the partition that minimizes $f(X)$ is the optimal solution. We here explain other terms of this function.

The first term, $b(X)$, is a Boolean function on X, formulated as

$$b(X) = \sum_{x \in X} b(x),$$

where

$$b(x) = \begin{cases} 0 & \text{if } |x| \leq S_{\max} \\ & \text{and the vertex-induced subgraph of } G \text{ by } x \text{ is connected,} \\ \infty & \text{otherwise.} \end{cases}$$

The second term, $h_d(X)$, evaluates the density of each predicted cluster in the following way:

$$h_d(X) = -\sum_{x \in X} density(x),$$

where

$$density(x) = \frac{w(x)}{\sqrt{|x|}}.$$

The last term, $h_{pn}(X)$, controls the number of proteins within the clusters of size two or more in X, which is denoted by $s(X)$,

$$h_{pn}(X) = (s(X) - \lambda)^2,$$

where

$$s(X) = \left| \bigcup_{x \in X \text{ s.t. } |x| \geq 2} x \right|$$

and λ is a user-specified target value of $s(X)$.

The whole scoring function, $f(X)$, of PPSampler2 is optimized by a Markov chain Monte Carlo algorithm. Details of the method are found in [1]. In performance comparison with the WI-PHI PPI data set [3], PPSampler2 is reported to outperform other methods, MCL [6], MCODE [7], DPClus [8], CMC [9], RRW [10], NWE [11], and PPSampler1 [1].

4 Overlapping Complexes

Overlapping complexes means different complexes which share the same at least one protein. Table 1 shows the number of pairs of overlapping complexes with the same number of shared proteins. We can see that the most frequent pattern is the case where one protein is shared by two protein complexes.

Any overlapping complexes cannot be found by a single execution of PPSampler2 because the structure of an output of PPSampler2 is a *partition* of proteins. However, the outputs of executions of PPSampler2 are not guaranteed to be the same because PPSampler2 is a sampling-based approach. Thus, a naïve approach for predicting overlapping complexes is to repeatedly execute PPSampler2 and merge all predicted complexes. Based on this idea, our method, ReSAPP (Repeated Simulated Annealing of Partitions of Proteins) [12] is devised. Briefly speaking, predictability of ReSAPP is reported to be slightly better than that of PPSampler2.

To the best of our knowledge, no direct modeling of overlapping complexes used in the main search process exists. Our next attempt to predict complexes including overlapping ones is to design a penalization term to find distinctive overlapping predicted complexes, which is formulated based on the Jaccard index as follows. For convenience, we denote by $m_{x,x'}$ the minimum size of $x, x' \subseteq V$, i.e., $m_{x,x'} = \min\{|x|, |x'|\}$. We define the dissimilarity score between x and x' as

$$h_{ds}(x, x') = \begin{cases} J(x, x') & \text{if } m_{x,x'} \leq 3 \text{ and } |x \cap x'| \leq 1, \\ & \text{or } m_{x,x'} \geq 4 \text{ and } \frac{|x \cap x'|}{m_{x,x'}} \leq \beta, \\ \infty & \text{otherwise.} \end{cases}$$

Namely, the criteria for the small clusters of size two and three and the larger ones are different. In the former case, x and x' are allowed to share only one protein. This constraint is reasonable due to their smallness. The latter case is formulated as

Table 1 The frequency of overlap sizes of protein complexes in CYC2008. The row of "#shared protein" shows the number of shared proteins by two complexes. The row of "Frequency" gives the number of pairs of overlapping complexes

#shared protein	1	2	3	4	5	6	7	8	9	10	11	12	13	14	15	16	17
Frequency	151	22	9	13	4	1	10	1	0	1	1	1	0	0	0	1	1

follows. If the ratio of the number of shared proteins to the minimum size among x and x' is less than or equal to β, the penalty is the Jaccard index, $J(x, x')$, and ∞ otherwise. Then, term $h_{ds}(X)$ is defined as follows:

$$h_{ds}(X) = \sum_{x,x' \in X} h_{ds}(x, x').$$

We added this term to the scoring function of PPSampler2. In addition to it, we made some modification of other terms of the scoring function. The format of an output in this new approach is changed to a set of subsets of V (no longer a partition of V). We also devise a new Markov chain Monte Carlo update procedure to propose overlapping predicted complexes. The resulting method is called RocSampler. Details of RocSampler will be found in the full version of [2].

We evaluated predicted clusters of proteins by a stringent matching criterion [13], which can be explained as follows. At first, x is said to match k with matching threshold η if $J(x, k) \geq \eta$. Let X be a set of all clusters predicted by a method, and K be a set of all known complexes. For subsets, $X' \subseteq X$ and $K' \subseteq K$, we formulate the two notations,

$$N_{pc}(X', K', \eta) = \{x | x \in X', \exists k \in K', J(x, k) \geq \eta\},$$
$$N_{kc}(X', K', \eta) = \{k | k \in K', \exists x \in X', J(x, k) \geq \eta\}.$$

Namely, X' is evaluated with K' and vice versa. For an integer i (≥ 2), we denote by $X|_i$ the subset of X whose elements are of size i, and by $X|_{\geq i}$ the subset of X whose elements are of size i or more. Similarly, $K|_i$ and $K|_{\geq i}$ for K are defined. The *precision* and *recall* are defined as follows:

$$\begin{aligned} & precision(X, K) \\ = & \frac{|N_{pc}(X|_2, K|_2, 1)| + |N_{pc}(X|_3, K|_3, 1)| + |N_{pc}(X|_{\geq 4}, K|_{\geq 4}, 0.5)|}{|X|}, \\ & recall(X, K) \\ = & \frac{|N_{kc}(X|_2, K|_2, 1)| + |N_{kc}(X|_3, K|_3, 1)| + |N_{kc}(X|_{\geq 4}, K|_{\geq 4}, 0.5)|}{|K|}. \end{aligned}$$

Namely, predicted clusters and known complexes of size two and three are evaluated with the exact match criterion, and the remaining larger ones are evaluated with $\eta = 0.5$. The *F-measure* of X to K is the harmonic mean of the corresponding precision and recall, i.e.,

$$F(X, K) = 2 \cdot \frac{precision(X, K) \cdot recall(X, K)}{precision(X, K) + recall(X, K)}.$$

The F-measure score of RocSampler is 0.44, followed by 0.37 (PPSampler2 [1]), 0.34 (ClusterONE [14]), and 0.31 (SPICi [15]). Thus, we can say that RocSampler outperforms the others.

Acknowledgements This work was partially supported by JSPS KAKENHI Grant Number 26330330.

References

1. C.K. Widita, O. Maruyama, PPSampler2: predicting protein complexes more accurately and efficiently by sampling. BMC Syst. Biol. **7**(Suppl 6), S14 (2013)
2. O. Maruyama, Y. Kuwahara, Rocsampler: regularizing overlapping protein complexes in protein-protein interaction networks, in *Proceeding of 2016 IEEE 6th International Conference on Computational Advances in Bio and Medical Sciences (ICCABS)* (IEEE, 2016). (The full version of this work is sumitted to a journal)
3. L. Kiemer, S. Costa, M. Ueffing, G. Cesareni, WI-PHI: a weighted yeast interactome enriched for direct physical interactions. Proteomics **7**, 932–943 (2007)
4. S. Pu, J. Wong, B. Turner, E. Cho, S.J. Wodak, Up-to-date catalogues of yeast protein complexes. Nucleic Acids Res. **37**, 825–831 (2009)
5. A. Clauset, C.R. Shalizi, M.E.J. Newman, Power-law distributions in empirical data. SIAM Rev. **51**, 661–703 (2009)
6. A.J. Enright, S. Van Dongen, C.A. Ouzounis, An efficient algorithm for large-scale detection of protein families. Nucleic Acids Res. **30**, 1575–1584 (2002)
7. G.D. Bader, C.W.V. Hogue, An automated method for finding molecular complexes in large protein interaction networks. BMC Bioinform. **4**, 2 (2003)
8. M. Altaf-Ul-Amin, Y. Shinbo, K. Mihara, K. Kurokawa, S. Kanaya, Development and implementation of an algorithm for detection of protein complexes in large interaction networks. BMC Bioinform. **7**, 207 (2006)
9. G. Liu, L. Wong, H.N. Chua, Complex discovery from weighted PPI networks. Bioinformatics **25**, 1891–1897 (2009)
10. K. Macropol, T. Can, A.K. Singh, RRW: repeated random walks on genome-scale protein networks for local cluster discovery. BMC Bioinform. **10**, 283 (2009)
11. O. Maruyama, A. Chihara, NWE: node-weighted expansion for protein complex prediction using random walk distances. Proteome Sci. **9**(Suppl 1), S14 (2011)
12. S. Kobiki, O. Maruyama, ReSAPP: predicting overlapping protein complexes by merging multiple-sampled partitions of proteins. J. Bioinform. Comput. Biol. **12**(6), 1442004 (2014)
13. C.H. Yong, L. Wong, From the static interactome to dynamic protein complexes: three challenges. J. Bioinform. Comput. Biol. **13**, 1571001 (2015)
14. T. Nepusz, Y. Haiyuan, A. Paccanaro, Detecting overlapping protein complexes in protein-protein interaction networks. Nat. Methods **9**, 471–472 (2012)
15. P. Jiang, M. Singh, SPICi: a fast clustering algorithm for large biological networks. Bioinformatics **26**, 1105–1111 (2010)

An Overview of Methods to Identify and Manage Uncertainty for Modelling Problems in the Water–Environment–Agriculture Cross-Sector

A. J. Jakeman and J. D. Jakeman

Abstract Uncertainty pervades the representation of systems in the water–environment–agriculture cross-sector. Successful methods to address uncertainties have largely focused on standard mathematical formulations of biophysical processes in a single sector, such as partial or ordinary differential equations. More attention to integrated models of such systems is warranted. Model components representing the different sectors of an integrated model can have less standard, and different, formulations to one another, as well as different levels of epistemic knowledge and data informativeness. Thus, uncertainty is not only pervasive but also crosses boundaries and propagates between system components. Uncertainty assessment (UA) cries out for more eclectic treatment in these circumstances, some of it being more qualitative and empirical. Here, we discuss the various sources of uncertainty in such a cross-sectoral setting and ways to assess and manage them. We have outlined a fast-growing set of methodologies, particularly in the computational mathematics literature on uncertainty quantification (UQ), that seem highly pertinent for uncertainty assessment. There appears to be considerable scope for advancing UA by integrating relevant UQ techniques into cross-sectoral problem applications. Of course this will entail considerable collaboration between domain specialists who often take first ownership of the problem and computational methods experts.

Keywords Mathematics-for-Industry · Water resources
Uncertainty assessment · Uncertainty quantification

A. J. Jakeman
Australian National University, Canberra, ACT, Australia
e-mail: tony.jakeman@anu.edu.au

J. D. Jakeman (✉)
Sandia National Laboratories, Albuquerque, NM, USA
e-mail: jdjakem@sandia.gov

R. S. Anderssen et al. (eds.), *Agriculture as a Metaphor for Creativity in All Human Endeavors*, Mathematics for Industry 28,
https://doi.org/10.1007/978-981-10-7811-8_15

1 Introduction

The discipline of hydrology in particular continues to be very successful at dealing with single-issue water resource management problems and developing a wide array of techniques that can be applied by consultants to infrastructure building, protection, operational and planning issues. Australian Rainfall and Runoff [6], for example, is the latest compendium in a series that presents such techniques in the Australian context for assessing infrastructure such as roads, rail, airports, bridges, dams, stormwater and sewer systems, town planning, mining, flood management planning for urban and rural communities, flood warnings and flood emergency management, operation of regulated river systems and prediction of extreme flood levels.

In the water–environment–agriculture sector, there remain, however, more complex, multifaceted problems requiring a more sophisticated approach where surface and subsurface hydrologic modelling is now contributing. Water is in many parts of the world a resource that is heavily contested for irrigated agriculture, environmental/ecological and human demands including livelihoods. And it is a resource that may be subject to deteriorating quality that has exacerbating long-term effects, beyond those of limited supply, for drinking water, agricultural and aquatic ecological systems. Perhaps, the most serious long-term water resource issue in this respect is that of the declining quantity and quality of groundwater, which is especially severe in India, Pakistan, China and the USA (see [50]). In Australia, for example, the Murray–Darling Basin Plan of 2012 [mdba.gov.au] has generated much debate about Sustainable Diversion Limits that have been set, being largely for irrigation uses versus the share of water that is being devoted to the environment. Decisions have, however, been contested, not just because of uncertainty in the hydrological modelling but also lack of knowledge in ecosystem response, and inadequate attention to socioeconomic consequences for farmers and wider communities dependent on agriculture. For any such multifaceted water resource issue, there is also the inherent subjectiveness in how to weight multiple objectives, and the need to anticipate future changes in so-called nexus conditions such as climate and conflicts with other resource issues like energy development projects and food production. But with such complexity also comes opportunity where innovations in water storage, supply and delivery can be explored to improve water management outcomes. Such opportunities will, however, need to be explored taking due account of uncertainties.

Consequently, in many situations, water resource decision-making is a wicked problem because there is no optimal solution, uncertainty is pervasive and stakes are contested. A related concept with increasing currency is that of deep uncertainty defined by there being 'Fundamental disagreement about the driving forces that will shape the future, the probability distributions used to represent uncertainty and key variables, and how to value alternative outcomes' [60]. Uncertainty Assessment (UA), and as far as possible Uncertainty Quantification (UQ), is essential in many situations for making credible predictions and defensible decisions. The importance of UQ has been repeatedly established in many US Department of Energy white papers, such as those resulting from the ASCR scientific grand challenge workshop series between 2008 and 2010. On the other hand for some purposes, a qualitative grasp of uncertainties may suffice or be more appropriate, such as when stakeholders are learning together about a resource issue so that in the longer term decisions can be made on a more informed and less contested basis. In supporting decision-making, however, UA will require quantitative and qualitative components, and perhaps some empiricism in methods when assumptions do not possess nice properties.

Here, we present an overview of methods and concepts to address the identification and management of uncertainty for supporting mutual learning by stakeholders, for making more useful predictions and/or for informing decisions, with a focus in the water–environment–agriculture resource management cross-sector. But the methods apply more widely. The motivation for this paper has been stimulated by the fact that there has been a lot of development of models of environmental processes, much less (but growing) generation of information about uncertainty and sensitivity in those models, very little frank reporting of uncertainties and very little trust in, or explicit use of, uncertainty information by users of models [49]. This is especially the case for issues in the water–environment–agriculture cross-sector. We use groundwater models and their links to other sectors to exemplify many of the points raised in the paper. Resulting cross-sectoral models are often called integrated (environmental) models.

2 Groundwater Models

There are many models used to simulate groundwater flow through an aquifer. Here, we choose to focus on the change in the hydraulic head h in a (1D–3D) spatial domain Ω over some time T using the following partial differential equation (PDE):

$$S_s \frac{\partial h(\boldsymbol{x}, t, z)}{\partial t} = -\nabla \cdot (-K(\boldsymbol{x}, z)\nabla h(\boldsymbol{x}, t, z)) - G(\boldsymbol{x}, t, z), \qquad (1)$$
$$\boldsymbol{x} \in \Omega, \ t \in [0, T], \ z \in I_z,$$

subject to some boundary conditions on $\partial\Omega$. Here, K is the hydraulic conductivity, S_s is the specific storage or storativity, G represents any source terms, such as pumps and wells and recharge, and $z = (z_1, \ldots, z_d)^T \in I_z$ are variable model parameters.

3 Sources of Uncertainty

We begin with a concise list of places where uncertainty arises in the modelling and decision support process. The list is fairly generic in that it should apply to problems characterised by deep uncertainty, not just those in the water resource and related sectors. Sources of uncertainty that need to be considered and managed derive from:

3.1 Data

Data are imprecise, often sparse in space and/or time, with systematic and or random errors and/or inadequate coverage of conditions. These errors affect calibration of the model while data input errors also affect outputs when using the model in a predictive or simulation mode.

Take our example of a 2D or 3D partial differential equation model to predict groundwater levels in an aquifer (1). The major parameters in such a model are the spatially varying conductivity (K) and storativities S_s. One must estimate such parameters largely from water level data obtained where bores are installed and water levels measured. This inverse problem is ill-posed in part due to limited amounts of data and measurement errors. Of course in practice such data have large errors partly because they are measured at coarse time intervals and at a relatively modest number of locations.

3.2 Future Forcing Conditions

There are many unknown variables or states of the model such as climate, demography, prices and sectoral and cross-sectoral policy changes that affect an integrated model of water and land management. Some of these unknowns can be modelled through the source term G in (1). Some apply to other sectoral component models.

In an aquifer context, future climate will affect the amount of recharge of precipitation to the groundwater, making predictions uncertain. Cross-sectoral issues creating future uncertainties may relate to the interactions of proposed energy extraction projects with existing groundwater uses for agriculture, or a government policy to issue more groundwater access to increase food production.

One way to manage such uncertainty is through exploratory scenario modelling where one hypothesises such uncertain futures and investigates the associated model outputs of interest (e.g. [59] so as to explore a range of outcomes, perhaps assessing how robust outputs are to future changes or even shocks [61]).

3.3 Parameters and/or Initial/Boundary Conditions

Estimated parameters will always have uncertainty but so will parameters that are considered known or can be measured. For example, in the latter case aquifer properties vary across very small scales yet a parameter value for K obtained from a pump test at a specific location may be used to represent them at some specified larger scale.

The chosen model parameterisation can have significant effect on whether a model can reproduce experimental observations. This is particularly true for parameterisation of spatially or temporally varying fields such as conductivity. The complexity of the parameterisation of conductivity can range from a single parameter for a homogeneous aquifer, to multiple parameters for a regional conductivity, to $O(10^5 - 10^6)$ parameters for a fully spatially distributed conductivity. A single parameter may be easily estimated from data, and however may result in poor fitting to data, whereas a highly distributed conductivity may lead to over-fitting and only a subset of parameters being informed by data.

3.4 Prior Knowledge

Prior knowledge may be used to constrain parameters in the formulated model structure. Inappropriate constraints may underestimate or overestimate uncertainty such as the way priors are selected for estimating aquifer parameters for conductivity and storativity. For example, an underestimation of the variance in model priors will lead to a misleading underestimation in the uncertainty of outputs of a groundwater model. Similarly, an over-estimation of prior uncertainty can lead to overly conservative estimates of uncertainty in predictions.

3.5 Model Formulation

All models are only approximations of reality. Uncertainties are introduced by the mathematical expressions chosen to describe the relationships between system variables and components. The same processes can be modelled using vastly different modelling methodologies, such as empirical, statistical, physical or hybrids of these approaches. Once a model structure has been chosen, it must then be implemented. For example, a finite element method may be used to solve a system of partial differential equations. The numerical method used to implement a given model and the type and level of discretisation are additional sources of model structure uncertainty.

Some aspects of model uncertainty may be known but non-identifiable in that there are multiple sets of parameter values that explain the model output. This non-uniqueness may apply to the model itself, even with exact/ideal data (a priori non-identifiability) or due to data being insufficiently informative (a posteriori non-identifiability).

3.6 Model Purpose and/or Objectives

It is crucial that these address the real issues, a problem often being that there has not been a thorough investigative and engagement process to identify the issues at stake and the form of advice needed, due for instance to resource limitations, lack of experts and perspectives, oversimplification of the issues or lopsided scientist push. But sometimes it can be just that the modeller does not relate the aims of the modelling to either the objective functions used to optimise model parameters or relevant performance measures [9].

As an example, one may wish to accurately predict the level of the aquifer at a set of specific locations. In this case, a very fine-scale spatial model will be important for capturing the desired quantities. However, a lumped model, which may be good at predicting total water volume in the aquifer, would not be capable of predicting local quantities accurately.

3.7 Verification/Validation Process

If this step is inadequate, there may be an overconfidence in the model's capacity and limitations. In any case, the less comprehensive this step, the less certain the model results. A common glaring deficiency is the omission of a cross-correlation analysis between model residuals (predictions minus corresponding observations) and model inputs to assess if there seems to be something missing in the models explanation of outputs. Verification and validation (V&V) must not be carried out deterministically but rather executed to account for the model uncertainty, e.g. variation in convergence rates of mesh refinement studies, due to parameter uncertainties.

3.8 Communication Process

There is often a disconnect between decision-makers and the people undertaking modelling and simulation. The metrics presented to decision-makers can be too complex and the amount of data overwhelming. The communication process needs to be iterative with the modellers providing information on what is most scientifically relevant and the decision-makers communicating what is important to them [48].

3.9 Socio-Environmental and Hydro-Ecological Systems

These system types present especially difficult challenges for uncertainty assessment and management. Model components of such systems are uncertain for most, if not all, of the reasons above. Uncertainties propagate between components, and propagation is not nicely dealt with in an analytic framework due to a mix of model types, results being contested and multiple outputs needing to be weighted in different ways to try and satisfy stakeholders possessing different values. In the latter case, the Murray–Darling Basin Plan process is illustrative. Thus, irrigators tend to have interests in having sufficient water for their needs in as many seasons as possible while city dwellers may feel that attending to the ecological needs of the river system and its groundwaters should involve a precautionary approach.

4 Methods and Concepts to Address Uncertainty

Despite the numerous challenges faced when attempting to understand the effects of model uncertainties, there are numerous existing methods and concepts that can be invoked in the quest for enhancing mutual learning among groups about a problem, better-informing predictions and/or supporting decisions under uncertainty. In this section, we provide a list of useful methods for exploring model uncertainty.

For many if not all the methods below, uncertainty analysis can be enhanced by simplifying the problem for a useful outcome by narrowing or exacting the purpose. It is always a judicious first step to be specific about objectives/modelling purpose [49].

At the highest level, one can ask is the objective strictly among the following: increasing understanding; prediction; discriminating qualitatively among decision options and their outcomes (such as predicting the direction of changes between two management options); forecasting a few time steps ahead or social learning. Social learning for instance does not demand as much knowledge about uncertainty, as say prediction, because it may be a way of using modelling to share knowledge and perceptions about a problem to generate trust or even consensus. Forecasting may also not require as accurate a model as prediction because one can use recent observations to update a model forecast. Discriminating among decision options may merely require one to be confident that one option is more likely to be better than another under certain circumstances. Prediction, however, may require more precision but one should be aware of the precision needed as it will inform the method and knowledge needed to achieve it. [62] for example considers a range of paradigms to undertake integrated modelling of socio-environmental systems including Bayesian networks, system dynamics, agent-based models and knowledge-based models. They give guidance on the appropriate paradigm according to modelling purpose, breadth of issues being considered, level of spatio-temporal detail required, nature of the data (quantitative only or also qualitative) and ease of representing uncertainty.

Consideration of the form and scale of model output is particularly relevant in the face of deep uncertainty. It may be for example that one does not need a time series for each of the various outputs but perhaps a probability distribution or some low-order moments. Consider that the output required is information about the interactions between surface (river) and groundwater (aquifer) at some spatial scale. A daily time series may be needed if one is interested in ecological impacts on a species but for other purposes such as water accounting it may be that just a flux is required over some longer term period, say seasonally.

In the following, we will assume that a set of n_q outputs $\boldsymbol{q} = (q_1, \dots, q_{n_q})^T$ of the model have been identified as relevant to the modelling objectives. For groundwater modelling using (1), these quantities of interest (QoI), $q_i(\boldsymbol{z})$, $i = 1, \dots, n_q$, are linear or nonlinear functionals of the PDE solution. Examples of QoI include the temporal and spatial average of hydraulic head within the aquifer or the minimum hydraulic head at a point $\boldsymbol{x}^\star \in \Omega$ over the life of the simulation,

$$q(z) = \int_0^T \int_\Omega h(\boldsymbol{x}, t, z)\, d\boldsymbol{x}\, dt \quad \text{and} \quad q(z) = \min_{t\in[0,T]} h(\boldsymbol{x}^\star, t, z),$$

respectively.

In addition to exacting the purpose of the deterministic model, one must also consider the purpose of the uncertainty analysis. For example, does one want to identify important parameters, predict mean behaviour or quantify risk. Quadrature methods are ideal for computing expectation of model outputs but cannot be used for estimating probability distributions.

Quantifying risk, that is, providing quantitative measures of risk associated with an event, is often a desired outcome of uncertainty analysis. Various measures can be employed to quantify risk and can be selected based upon a modeller's or decision-maker's aversion to or acceptance of risk. The most common risk measure is standard deviation, but other measures often used in the financial industry are value at risk and conditional value at risk. Quantifying risk can be used to understand the probability of rare events that may have catastrophic consequences.

It may also pay to consider concepts of vulnerability, risk and/or robustness when attempting to get a more definitive handle on uncertainties. These terms can apply to the formulation of the objectives/outputs of interest or they can apply to scenario analysis where one identifies uncertainties in inputs and parameters which lead to vulnerable, risky or robust outcomes.

4.1 *Quadrature*

Quadrature is often used in uncertainty analysis to compute moments of QoI, for example, variance used to quantify risk. A quadrature rule consists of a set of points $z^{(m)}$ and a corresponding set of weights $w_m, m = 1, \ldots, M$, which are used to numerically approximate integrals via

$$\int_{I_z} f(z)\, d\mu(z) \approx \sum_{m=1}^{M} w_m f\left(z^{(m)}\right). \tag{2}$$

for some measure μ. The quadrature is usually constructed so that it is exact for some function space. The most common choice is the space of total-degree polynomials of a given degree.

There are numerous approaches for computing polynomial-based quadrature rules. For univariate functions, Gaussian quadrature is the most commonly used approach [27]. Gaussian quadrature rules are derived from the roots of the polynomials orthogonal to the measure μ. The resulting quadrature weights are always positive, and the rules are optimal in the sense that, for a Gaussian quadrature rule of degree of exactness p, no rule with less points can be used to integrate any p-degree polynomial.

When integrating multivariate functions with respect to tensor product measures on a hypercube, accurate and efficient quadrature rules can be found by taking tensor products of one-dimensional Gaussian quadrature rules. These rules will be optimal, in the aforementioned sense, for functions that can be represented exactly by tensor product polynomial spaces of degree p. However, the use of such quadrature rules is limited to a small number of dimensions, say 3–4, because the number of the points in the rule grows exponentially with dimension.

Sparse grid quadrature methods [29, 84] have been successfully used as an alternative to tensor product quadrature for multivariate functions. The number of points in sparse grid rules only grows logarithmically with dimension. Unlike tensor product rules, however, the quadrature weights will not all be positive. Sparse grid quadrature delays the curse of dimensionality, by focusing on integrating polynomial spaces that have high-degree univariate terms and multivariate terms for which the degree of the term decreases quickly with the interaction order of the term, i.e. the number of non-constant univariate polynomials that make up that term. The exact rate of decay of the importance of an interaction depends on the exact type of sparse grid method being used.

High-dimensional cubature rules can often be more effective than sparse grid rules when integrating functions that are well represented by total-degree polynomials. These rules have positive weights and typically consist of a very small number of points. However, such highly effective cubature rules are difficult to construct and have only been derived for a specific set of measures, integration domains and polynomial degree of exactness [89, 100].

When a function is well approximated by polynomials, the aforementioned polynomial quadrature methods will provide an accurate approximation of the integrals in (2). Specifically, the error in the estimate will decay exponentially with the number of points M. The exact rate of convergence depends on the type of method employed. The convergence rate of tensor product and total-degree rules with M points is $O(N^{-r/n})$ for a function with continuous $r\text{-}th$ partial derivatives.

When the number of variables x is sufficiently high, all the aforementioned quadrature methods will produce quadrature rules with an intractably high number of points. In this setting, Monte Carlo (MC) and Quasi-Monte Carlo (QMC) methods provide a more tractable means of integrating high-dimensional functions. Unlike the polynomial-based quadrature rules which consist of varying quadrature weights in (2), MC and QMC methods assign equal weights $1/M$ to each point. MC methods select the quadrature weights randomly from the probability measure μ. The convergence rate of MC methods is $O(\text{Var}_z[q(z)]M^{-1/2})$. In contrast, QMC methods [40, 86] have a convergence rate of $O(M^{-1}\log(M)^d))$ and select quadrature points that minimise the discrepancy of the quadrature points. Conceptually, discrepancy can be thought of as a measure of the distance between sets of points. Unlike MC methods, which can be used with any probability measure, QMC methods are designed for the

Lebesgue measure and transformations must be applied to the points to make QMC methods applicable to other probability measures.

4.2 Importance Sampling

One often wants to compute the probability of a rare event, for example, groundwater level dropping below a certain level δ. The probability of the rare event can be computed using Monte Carlo sampling

$$P_f = \text{Prob}[q((z)) < \delta] \approx \frac{1}{M}\sum_{m=1}^{M} \mathbb{I}_{\{q(z)<\delta\}}(z^{(m)}), \quad \mathbb{I}_A(z) = \begin{cases} 1, & z \in A \\ 0, & z \notin A \end{cases} \tag{3}$$

where $z^{(m)}$ are a set of samples drawn from the probability distribution of the random variables z. Due to its poor convergence rate (see Sect. 4.1), Monte Carlo sampling quickly becomes infeasible as the probability of failure P_f decreases. If $P_f = O(10^{-4})$, MC sampling will require $O(10^8)$ samples.

An effective alternative method to compute the probability of rare events is importance sampling (IS) [13, 88]. Instead of sampling from the probability distribution $\pi(z)$ of z, IS introduces a biasing distribution $\hat{\pi}(z)$ and rewrites (3) as

$$P_f \approx \frac{1}{M}\sum_{m=1}^{M} \frac{\pi(z)}{\hat{\pi}(z)} \mathbb{I}_{\{q(z)<\delta\}}(z^{(m)}),$$

where $z^{(m)}$ are now drawn from the biasing distribution $\hat{\pi}(z)$. The efficiency of IS is dictated by choice of the biasing distribution, which directly controls the amount of samples that fall within the desired rare event domain. The construction of a good biasing distribution is a difficult task and is the main difference between IS methods.

It is worth remarking that a number of other non-sampling-based methods have been developed for computing the probability of rare events. Some of the most popular methods include the first- and second-order reliability methods (FORM and SORM) [43, 58]. These methods are typically much less expensive than sampling-based methods but have reduced and limited accuracy.

4.3 Density Estimation

Density estimation (DE) [83, 94] is sometimes used to gain more detailed insight into the uncertainty associated with a prediction than can be obtained from simply estimating moments or probability or rare events [87]. DE is also sometimes necessary when an uncertain prediction of one model is an input to another model [92].

Given a set $z^{(i)}$ of samples drawn from an unknown distribution with unknown probability density function f, DE constructs an estimated density function $\hat{f}$ based upon the data. Density estimation can be classified into two broad classes, parametric and non-parametric. Parametric methods assume the form of the distribution and use data to estimate the parameters (e.g. the mean and variance of a normal distribution) of the assumed functional form, whereas non-parametric methods do not assume a parametric form but rather some less restrictive assumptions on density smoothness.

The non-parametric method of kernel density estimation (KDE) has become one of the most popular DE techniques. Gaussian KDE approximates the true density as a linear combination of squared-exponential kernels centred at each of the data points

$$\hat{f}(z) = \frac{1}{nh}\sum_{i=1}^{n} K\Big(\frac{z - z^{(i)}}{h}\Big).$$

The accuracy of Gaussian KDE is dependent on the selection of the kernel bandwidth h. The most simple methods for choosing the bandwidth are based upon asymptotic arguments. Such kernel density estimation schemes are known to converge $\mathcal{O}(M^{-4/(4+d)})$ in the mean-squared error [93] and $\mathcal{O}(M^{-2/(4+d)})$ in the L_1-error [25]. More computationally expensive methods, such as ones based upon cross-validation, can also be used to achieve increased accuracy for the same number of samples M. See [56] for a survey of bandwidth selection algorithms. Adaptively selecting bandwidths for each kernel can result in further improvement in accuracy [39, 65]. Other non-parametric density estimation techniques worth mentioning include spline-based density estimation [42] and methods that include so-called soft information, such as shape, support, continuity, smoothness, slope, moments, etc. to improve the accuracy of the density estimate [79].

4.4 Sensitivity Analysis

Sensitivity analysis (SA) is used to identify the sources of uncertainty and their relative influence [81]. Specifically, SA can be used to determine which input factors contribute most to the output variability; the presence and types of interactions between input factors; and the presence of nonlinearities within the model. Identifying important inputs allows future research to focus on increasing knowledge of the behaviour of the inputs in order to constrain the input variability and hence reduce the output uncertainty. Identifying insignificant inputs can also help refine model structure through the removal of parameters that have negligible effect on the behaviour of the model. Some commonly used SA techniques include local sensitivity methods, variance-based techniques and regional sensitivity analysis.

Local SA methods, such as automatic differentiation [97] and the Morris method [67], characterise sensitivity by partial derivatives or gradients at the local point. These methods are generally very simple and easy to implement and work well for linear models. However, when the model is nonlinear, the results obtained at a nominal point are in general not representative of the entire space. Variance-based techniques, such as the Fourier Amplitude Sensitivity Test (FAST) [23, 80] and the Sobol method [85], involve decomposing the output variance into parts attributed to individual variables and interactions between variables. Regional Sensitivity Analysis (RSA) [45] partitions model realisations into behavioural sets and non-behavioural sets, that is, the set of input factors that satisfy the problem constraints and those that do not.

Recently, a new technique known as active subspaces has become popular for identifying lower dimensional structure. Unlike the aforementioned methods, active subspaces can identify directions in parameter space which may not be aligned with the parameter axes that significantly influence a QoI. These directions are the eigenvectors of a matrix derived from the gradient of the parameter-QoI map [55].

Related to sensitivity analysis is break-even analysis. Break-even analysis identifies model variables at tipping points where one is considering management options two at a time and conditions, and uncertainties can be generated to define at which points one option is as good as another [37].

4.5 Deterministic Calibration and Bayesian Inference

When observational data are available, these should be used to inform prior assumptions of model uncertainties. This so-called inverse problem that seeks to estimate uncertain parameters from measurements or observations is usually ill-posed. Many different realisations of parameter values may be consistent with the data. The lack of a unique solution can be due to the non-convexity of the parameter-to-QoI map, lack of data, and model structure and measurement errors.

Deterministic model calibration is an inverse problem that seeks to find a single-parameter set that minimises the misfit between the measurements and model predictions. A unique solution is found by simultaneously minimising the misfit and a regularisation term which penalises certain characteristics of the model parameters.

In the presence of uncertainty, we typically do not want a single optimal solution, but rather a probabilistic description of the extent to which different realisations of parameters are consistent with the observations. Bayesian inference [57] can be used to define a posterior density for the model parameters z given observational data $\boldsymbol{y} = (y_1, \ldots, y_{n_y})$:

$$\pi_{\text{post}}(z) = \frac{\pi(\boldsymbol{y}|z)\pi(z)}{\int_{I_z} \pi(\boldsymbol{y}|z)\pi(z)dz}, \tag{4}$$

where any prior knowledge on the model parameters is captured through the prior density $\pi(z)$. The discrepancy between the observations and the simulation model is encoded in the likelihood $\pi(\boldsymbol{y}|z)$ and should reflect the true nature of this discrepancy. The true form of the error is often unknown, and it is often assumed that the deviation between the model prediction and the model errors is given by additive Gaussian errors with mean zero and covariance $\boldsymbol{\Sigma}_{\text{noise}}$, that is

$$y = q(z) + \epsilon, \quad \epsilon \sim N(\mathbf{0}, \boldsymbol{\Sigma}_{\text{noise}}).$$

Typically, a functional representation of the posterior distribution π_{post} is not constructed, but rather samples from the posterior are obtained using Markov chain Monte Carlo (MCMC) sampling methods [33]. Although MCMC sampling is the most popular and efficient means of sampling from a posterior density, the convergence rate of MCMC sampling degrades with parameter dimension. The decreased rate of convergence is due to increased mixing time, which produces higher variance in posterior estimates [66, 78]. However, the performance of MCMC can be improved significantly in higher dimension through dimension reduction and leveraging gradient and Hessian information of the parameter-to-QoI map [14, 22]. These methods achieve greater efficiency by identifying lower dimensional subspaces that characterise the difference between the posterior distributions, induced by the observational data, relative to the prior and have been used to reduce the effective dimension of the input variables from $O(10^5) - O(10^6)$ to $O(10^1) - O(10^2)$.

There is a strong relationship between Bayesian inference and nonlinear least squares parameter estimation [90]. When the prior and error models are Gaussian with mean zero and covariance Σ_{pr} and Σ_{noise}, respectively, finding the maximum value of the posterior density (4) or rather the minimum of its negative logarithm (neglecting constants), i.e.

$$\min_{z \in I_z} \left(y - q(z)\right)^T \boldsymbol{\Sigma}_{\text{noise}} \left(y - q(z)\right) + z^T \boldsymbol{\Sigma}_{\text{pr}} z,$$

is equivalent to solving a regularised least squares problem where the regularisation term is the value of the negative logarithm of the prior density.

In situations where the proper Bayesian likelihood in (4) is not known or is prohibitively expensive to evaluate, one can use an alternative or misfit function that measures the discrepancy between a chosen set of statistics of model outputs and the corresponding estimates from the data. These methods are often referred to as likelihood-free methods and include techniques such as approximate Bayesian computation (ABC) [8] and generalised likelihood uncertainty estimation (GLUE) [10].

4.6 Response Surface and Emulation Methods

Many simulation models are extremely computationally expensive such that adequately understanding their behaviour and quantifying uncertainty can be computationally intractable for any of the aforementioned techniques. Various methods have been developed to produce surrogates of the model response to uncertain parameters. The most efficient are goal-oriented in nature and target very specific uncertainty measures. Generally speaking, surrogates are built using a "small" number of model simulations and are then substituted in place of the expensive simulation models in future analysis. Some of the most popular surrogate types include polynomial chaos expansions (PCE) [31, 99], Gaussian processes (GP) [76] and sparse grids (SG) [15].

For PCE, GP and SG, a number of unique methods can be used to build each surrogate. The most popular methods to construct PCE include pseudo-spectral projection [20], least squares [18] and compressive sensing [26, 53]. Maximum likelihood estimation and Bayesian inference are the two most popular approaches for building GP [76] and sparse grids can be built isotropically (treating all variables equally) [71], with dimension adaptivity [30] or local adaptivity [52, 74]. The choice of the best surrogate method is problem dependent [34].

Some major benefits of each method include analytic computation of mean and variance and Sobol sensitivity indices from a PCE; estimates of error in the surrogate when using GP; and the computational efficiency of constructing a sparse grid (not to be confused with efficiency in the number of samples required). Moreover, there are numerous approaches for building PCE, GP and SG.

Many existing studies that utilise surrogates do not consider the distribution of z when building the surrogate. Probabilistic information is then only included when sampling the surrogate to compute moments for example. If this approach is adopted for unbounded variables, such as Gaussian variables, the domain on which the surrogate is constructed must encapsulate the regions of high non-zero probability [51]. Not taking into account the probability of the random variables when building a surrogate often results in a loss of accuracy when compared to methods that leverage such information. Probability-unaware methods lose efficiency because to build a stable approximation one must often sample in regions of very low probability. Recently, there have been some attempts to design efficient sampling schemes for building GP and PCE for different random variable distributions. For example, see [41, 69] when building PCE using least squares and compressive sensing and [34] when building a GP.

Surrogates have been used successfully to reduce the number of model simulations to perform moment estimation of the parameter-QoI map [98], sensitivity analysis [91], importance sampling [63] and Bayesian inference [75]. However, the number of samples needed to build a surrogate increases with dimension, in many cases exponentially, which limits the use of surrogates to moderate dimensional problems or computationally inexpensive models. To address this challenge, surrogates are often built on a reduced set of important variables, which have been identified via sensitivity analysis [19].

4.7 Multi-fidelity Methods

Multi-fidelity methods attempt to reduce the computational burden of uncertainty analysis, by enriching a small number of high-fidelity simulations with a larger number of lower fidelity simulations. In groundwater models, the lower fidelity models may just represent coarser mesh discretisations, or a 2D and/or 1D approximation of a 3D model, an empirical model or even a surrogate (see Sect. 4.6). Multi-fidelity approaches enable more rapid convergence to high-fidelity statistics when such lower fidelity models provide predictive utility. In particular, multi-fidelity UQ can converge more rapidly than single-fidelity UQ in cases when there is a high correlation between predictions of the models of varying fidelity.

Many multi-fidelity methods, such as Multi-level Monte Carlo (MLMC) [32], Control Variate Monte Carlo (CVMC) [32] and surrogate-based methods [70], target the discrepancy between two levels of fidelity. Multi-level and control variate Monte Carlo sampling can efficiently estimate statistics of the QoI, such as $\mathbb{E}[q]$, by leveraging models with strong correlation. The RMSE of a Monte Carlo estimate of some function $g(z)$ is $\epsilon = M^{-1}\text{Var}[g]$ and if we can reduce the variance of g then we can achieve a desired level of accuracy in the estimator $\text{Var}[g]$ using less samples. As an example consider Control Variate Monte Carlo and again let q be the solution to the groundwater model in (1), and let f be another model (called a control variate) which is correlated with q, then we can compute an unbiased estimator for the mean of q using

$$\mathbb{E}_{\text{CV}}[q] = \frac{1}{M}\sum_{m=1}^{M}\left(q(z^{(m)}) - \lambda\left(f(z^{(m)}) - \mathbb{E}[f]\right)\right), \quad \lambda = \rho\sqrt{\frac{\text{Var}[q]}{\text{Var}[f]}},$$

where ρ is the correlation between q and f. Using the optimal value of λ given above, the variance of the control variate estimator $\mathbb{E}_{\text{CV}}[q]$ is a factor of $1-\rho^2$ smaller than the variance of the single-model estimator $\mathbb{E}[q]$. By reducing the variance of the estimator, we are able to reduce the error in the Monte Carlo estimate of the mean of q. Multi-level Monte Carlo works in a similar way by estimating the expectation of q by computing expectations of the discrepancy of pairs of models of increasing fidelity. For example, consider a two-level MLMC estimate of the expectation of q with a low-fidelity model q_0

$$\mathbb{E}_{\text{ML}}[q] = \mathbb{E}[q_0] + (\mathbb{E}[q] - \mathbb{E}[q_0]) = \frac{1}{M_0}\sum_{m=1}^{M_0} q_0(z^{(m)}) + \frac{1}{M_1}\sum_{m=1}^{M_1}\left(q(z^{(m)}) - q_0(z^{(m)})\right),$$

where, if the correlation between q_0 and q is high, then $M_1 \ll M_0$. MLMC and CVMC were developed independently but have now been successfully used together [28]. It is interesting to note that it was shown in [73] that surrogate models that inform different aspects of the high-fidelity model are better than surrogate models that are accurate but lack a rich diversity. Focusing on the discrepancy between models has also been used with methods that build surrogates of the discrepancy and exploit structure, such as sparsity or reduced spectral content of the discrepancy, which may not be present in the single-fidelity models.

Finally, we mention a vastly different approach for multi-fidelity uncertainty quantification [68] which uses a low-fidelity model of a PDE to choose a set of samples

z that correspond to a linearly independent set of low-fidelity PDE solutions h_0, and then evaluates the high-fidelity PDE at these samples and computes a multi-fidelity approximation of the solution

$$\hat{h}(x, t, z) = \sum_{i=1}^{m} \langle h_0(x, t, z), h(x, t, z^{(m)}) \rangle h(x, t, z^{(m)}). \tag{5}$$

To evaluate the coefficients of the high-fidelity basis, one must evaluate the lower fidelity model and then project the resulting solution onto the high-fidelity basis. If the low- and high-fidelity models are well correlated, then this will be a reasonable approximation of the coefficients of the high-fidelity basis. Although the low-fidelity model has a constant error, the bi-fidelity construction (5) provably achieves exponential error convergence as the number of high-fidelity simulations is increased. The number of high-fidelity solutions required to achieve a given accuracy will be dependent on the rank of the span solution space of the high-fidelity model, that is, the number of linearly independent solutions.

We note here that unlike MLMC and CVMV, the low-rank solution approach and the surrogate-based discrepancy approach can be used to not only compute moments but, like the surrogates discussed in Sect. 4.6, also be used to for computing PDFs and rare events and even Bayesian inference.

4.8 Scenario Modelling with and Without Formal Probabilities

In its broadest sense, this involves exploring multiple, plausible scenarios about future conditions, model structure and parameter values. It can be used for many purposes, such as to promote discussion and sharing of knowledge and perspectives and/or to search for those scenarios that lead to good, intermediate and poor outcomes. At its core are simulation of model drivers and parameter samples, and analysis of the model QoI. It can potentially utilise many of the methods above to achieve computational efficiencies required for models with slow runtimes, as often occurs with integrated models and groundwater models.

4.9 Crash Testing/Torturing the Model

This can be similar to scenario modelling in that attempts are made to see what model parameter sets, observation periods and other conditions establish limitations or invalidate the model. This can involve examining the performance of the model through time and/or space to assess inadequate performance.

Multiple models are also a useful way to explore uncertainties in model formulations. Different model candidates or perspectives can be used with the other tools mentioned above and below to understand sources of uncertainty. Various techniques such as Bayesian model selection can then be used to assess the strengths and weaknesses of each, and under which conditions each model is more suitable.

4.10 *Qualitative Information About Uncertainty*

A qualitative approach may be sufficient or at least a useful adjunct in some cases. The emphasis in this case may be on how the prediction was produced, and the limitations involved in doing so [36]. As covered in [36], one way of approaching this is through quality assurance of the modelling process [77] and its constituent assumptions, whilst another is to include qualitative judgements about the information and how it is produced [95].

4.11 *Understanding Your Data and Its Relationship to the Model*

Simple textbook analysis of data to reveal their uncertainties before modelling is under-practised or at least under-reported. There is a wealth of tools available to detect outliers, trends, implausible correlations, timing errors in model response and generally extract information from data. The value of simple plotting and visualisation should not be ignored.

Often collection of experimental data is expensive and only a limited amount of experimental data can be obtained. Not all experiments, however, provide the same amount of information about the processes they are helping inform. Consequently, it is important to design experiments in an optimal way, i.e. to choose some limited number of experimental data to maximise the value of each experiment. Optimal experimental design (OED), that is, using physical models to guide experiment selection, has been shown to drastically improve the cost-effectiveness of experimental designs for a variety of models based on ordinary differential equations [11], partial differential equations [44] and differential algebraic equations [7]. OED has been developed in both Bayesian and non-Bayesian settings [5, 16, 96]. When model observables are linear with respect to the model parameters, alphabetic optimality criteria are often used [38].

4.12 *Multidisciplinary Analysis*

Often modelling environmental systems involves the integration of numerous components from multiple disciplines. Uncertainty analysis of such integrated system models is often challenging when applied to the integrated system. Simulation of the integrated system may be computationally expensive limiting the exploration of model uncertainties and often individual model components are managed by different groups, with varying computational software and hardware, which can hinder unified and automated modelling.

Recently, a number of methods have been developed that decompose system uncertainty analysis into uncertainty analysis of individual components, which can then be combined efficiently to assess system-level uncertainties [1, 82]. This approach allows the vast majority of computational effort, that is, UA of individual components, to be carried out in parallel. The assembly of the system-level analysis can be carried out using the component-level analysis with much less computational effort.

An example of a multicomponent system is shown in Fig. 1. In this example, two models, for example, neighbouring groundwater models that share a boundary, are used to provide input to a higher level model, which may be a model of a bird ecosystem. In this example, the coupling variables $\boldsymbol{b}_1$ and $\boldsymbol{b}_2$ represent boundary conditions along the shared interface between the two groundwater models. The system consists of three models with random variables $\boldsymbol{\xi}, \boldsymbol{\eta}$ and $\boldsymbol{\theta}$. These variables may be unique to each model or some or all variables may be shared between models so that the variables of the coupled system are $z = \boldsymbol{\xi} \cup \boldsymbol{\eta} \cup \boldsymbol{\theta}$. Often only a small number of random variables are shared between models, thus performing UA on each individual component can be performed in a much lower dimensional setting. This facilitates the use of efficient lower dimensional methods such as surrogates, which may not be tractable when applied to the integrated system [17]. Furthermore, often only a lower dimensional subspace of the coupling variables significantly affects the outputs of F_1 and F_2, and consequently dimension reduction techniques can be used to further increase the tractability of decoupled analysis [3, 64].

A number of different approaches exist for decoupling the system components. The first approach uses Monte Carlo sampling combined with fixed-point iteration that, for a given realisation of the random variables, iteratively finds the values of the coupling variables that produce consistent solutions $\boldsymbol{f}_1$ and $\boldsymbol{f}_2$. The iteration starts with arbitrary value of $\boldsymbol{b}_1$ being passed as input to F_2, $\boldsymbol{f}_2$ and $\boldsymbol{b}_2$ are then computed, and $\boldsymbol{b}_2$ is passed as input to F_1 and this procedure is repeated until convergence. Surrogate approaches have also been used to decouple multisystem analysis. For example, [4, 17, 21] introduce variables y that represent the dependencies b_1 and b_2 on the random variables z. Surrogates are then built over the random variables z and $\boldsymbol{b}$ in the component-level analyses. An appropriate coupling algorithm can then be applied to the surrogates to find the values of the coupling variables. In [82], a likelihood-based approach was proposed to decouple feedback loops, reducing the problem to a system of only one way coupling, resulting in a so-called feedforward problem, thereby removing the need for fixed-point iteration.

When the multicomponent system consists of only forward couplings, measure transformations can be used to increase the efficiency of the system analysis [1]. Each single component analysis can be performed using a proposal distribution for each of the unknown model inputs which are dependent on the outputs of other models. The proposal distribution (light blue in Fig. 1) must encapsulate the regions

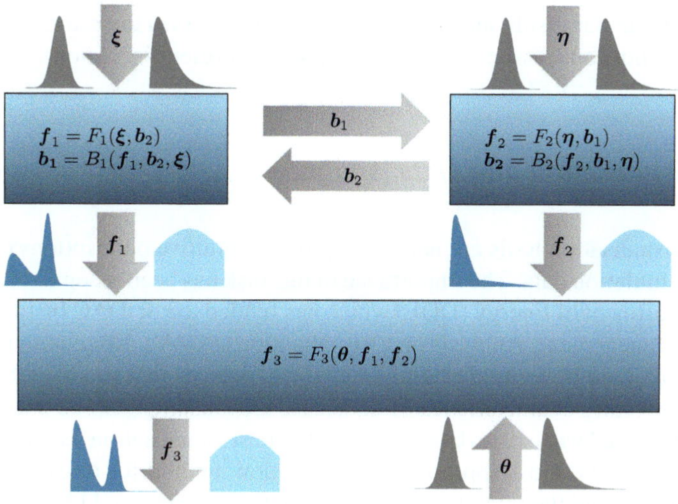

Fig. 1 An example of a multicomponent system model. The efficiency of uncertainty analysis of such systems can be increased by decoupling the individual model components, applying UA on each component and then reassembling the results to assess the coupled system. The distributions of the random variables z are shown in grey. Proposal distributions for the coupling variables (f_1, f_2) used in the decoupled analysis are shown in light blue. The true distributions of the coupling variables recovered during the assembly phase are shown in dark blue

of non-zero probability of the unknown distributions (dark blue). Samples of each model output can then be drawn from these proposal distributions for each model independently and in parallel.[1] For the lowest level models, there is no need for an approximation of the inputs as they are the known random variables. Starting with the output, f_1, f_2 of the lowest level models, importance sampling (at no extra simulation cost) is used to reweight the input to the next set of downstream components f_3 to obtain effective samples from the true distribution of f_3 (dark blue). The weights are simply the ratio of the true and known density of f_3 and the proposal density. These steps are repeated for each level of the system model hierarchy. The computation of these weights requires density estimation which can be expensive in higher dimensions (see Sect. 4.3). Consequently, optimisation approaches have been developed for reweighting samples without the need to estimate densities of model inputs and outputs [2].

In addition to algorithms that enable decoupled analysis of integrated systems, various software frameworks have been, and continue to be, developed to address (or in an attempt to address) the technical complexity and issues encountered when integrating separate models [24, 35, 72]. The approaches construct a (software) interface for each component model to allow interaction through a single formalised and unified framework. There have also been attempts to leverage software development technologies to provide an archive of a model used for a particular purpose or to

[1]In this example, we will assume that the coupling between F_1 and F_2 has been broken.

produce a particular result, and a common runtime platform to ensure differences in hardware do not cause issues (see [46, 47, 54] for further discussion).

4.13 Visualisation and Data Analysis

Effective visualisation tools are needed to provide intuitive descriptions of complex and large simulation data. The importance of this task has been recognised, including the US Department of Energy (DOE) which has funded the SciDAC Institute of Scalable Data Management, Analysis and Visualization (SDAV). Effective visualisation tools can facilitate better understanding of the processes that produce the data and identify interesting characteristics of data sets. Visualisation can be used to extract information from large-scale data; however, the best visualisation technique is problem dependent. The most appropriate visualisation method for a task is dependent on the data type and the intended goals of the data analysis. See [12] for a review of visualisation methods for uncertainty visualisation.

5 Conclusions

Hydrological models, of the surface and subsurface type, underpin modelling in the water resources sector. But many models in the environment sector, especially hydrological ones, are non-identifiable. Thus, analysis tools are essential to expose critical parameters so that either improved, identifiable (and likely simpler) model formulations can be obtained or the limitations of non-identifiability can be taken into account in uncertainty assessments.

In the paper, we have outlined a fast-growing set of methodologies, particularly in the computational mathematics literature, that are pertinent for uncertainty assessment. There seems considerable scope for advancing UA by integrating relevant ones into cross-sectoral applications. Of course, this will entail considerable collaboration between domain specialists who often take ownership of the problem and computational methods experts.

While it may seem obvious, the most basic step in UA is to ensure that the model addresses the questions being asked. Indeed, it must follow that one allows for the expense of essential analysis of model uncertainty and ascertains what uncertainties are crucial and concentrate on them. In this connection, attention to problem framing and stakeholder engagement is now generally regarded as crucial when the problem has deep uncertainty in order to manage it. Managing uncertainty has several elements including initially identifying and ranking the importance of its sources so that it can be reduced where possible and generally appreciated.

There continues to be too little crash testing of models to expose their limitations and too little reporting of those limitations and assumptions. A positive development has been the increasing use of sensitivity analysis to investigate major inputs and parametric contributors to the predictions from using a particular model structure and data set.

It is obvious that mutual learning about a problem can proceed in the face of uncertainty and the benefits of this process are not to be underestimated but it is also possible that good decision-making can also proceed in the face of uncertainty. Broadly, mathematical methods of modelling and analysis as well as clever, inclusive problem framing have much to offer to advance outcomes (understanding, social learning and decision support) for treating wicked problems in the water resource and related sectors.

Acknowledgements Sandia National Laboratories is a multimission laboratory managed and operated by National Technology and Engineering Solutions of Sandia, LLC., a wholly owned subsidiary of Honeywell International, Inc., for the U.S. Department of Energy's National Nuclear Security Administration under contract DE-NA-0003525. The article is a contribution to the NSF-funded National Socio-Environmental Synthesis Center project on Effective core practices for model-based integrated water resources management.

References

1. S. Amaral, D. Allaire, K. Willcox, A decomposition-based approach to uncertainty analysis of feed-forward multicomponent systems. Int. J. Numer. Methods Eng. **100**(13), 982–1005 (2014)
2. S. Amaral, D. Allaire, K. Willcox, Optimal l_2-norm empirical importance weights for the change of probability measure. Stat. Comput. **27**(3), 625–643 (2017)
3. M. Arnst, R. Ghanem, E. Phipps, J. Red-Horse, Dimension reduction in stochastic modeling of coupled problems. Int. J. Numer. Methods Eng. **92**(11), 940–968 (2012)
4. M. Arnst, R. Ghanem, E. Phipps, J. Red-Horse, Measure transformation and efficient quadrature in reduced-dimensional stochastic modeling of coupled problems. Int. J. Numer. Methods Eng. **92**(12), 1044–1080 (2012)
5. A.C. Atkinson, A.N. Donev. *Optimum Experimental Designs* (Oxford University Press, 1992)
6. J. Ball, M. Babister, R. Nathan, W. Weeks, P.E. Weinmann, M. Retallick, I. Testoni (eds.), *Australian Rainfall and Runoff: A Guide to Flood Estimation. Commonwealth of Australia* (Geoscience Australia, 2016)
7. I. Bauer, H.G. Bock, S. Krkel, J.P. Schlder, Numerical methods for optimum experimental design in DAE systems. J. Comput. Appl. Math. **120**(12), 1–25 (2000)
8. M.A. Beaumont, W. Zhang, D.J. Balding, Approximate bayesian computation in population genetics. Genetics **162**(4), 2025–2035 (2002)
9. N.D. Bennett, B.F.W. Croke, G. Guariso, J.H.A. Guillaume, S.H. Hamilton, A.J. Jakeman, S. Marsili-Libelli, L.T.H. Newham, J.P. Norton, C. Perrin, S.A. Pierce, B. Robson, R. Seppelt, A.A. Voinov, B.D. Fath, V. Andreassian, Characterising performance of environmental models. Env. Model. Softw. **40**, 1–20 (2013)
10. K. Beven, A. Binley, The future of distributed models: model calibration and uncertainty prediction. Hydrol. Process. **6**(3), 279–298 (1992)
11. H.G. Bock, S. Körkel, J.P. Schlöder, *Parameter Estimation and Optimum Experimental Design for Differential Equation Models* (Springer, Berlin, 2013), pp. 1–30
12. G.P. Bonneau, H.C. Hege, C.R. Johnson, M.M. Oliveira, K. Potter, P. Rheingans, T. Schultz, Overview and state-of-the-art of uncertainty visualization, in *Scientific Visualization: Uncertainty, Multifield, Biomedical, and Scalable Visualization*, ed. by C.D. Hansen, M. Chen, C.R. Johnson, A.E. Kaufman, H. Hagen (Springer, London, 2014), pp. 3–27

13. A. Bucklew, *Introduction to Rare Event Simulation* (Springer, 2004)
14. T. Bui-Thanh, O. Ghattas, J. Martin, G. Stadler, A computational framework for infinite-dimensional bayesian inverse problems part i: The linearized case, with application to global seismic inversion. SIAM J. Sci. Comput. **35**(6), A2494–A2523 (2013)
15. H.-J. Bungartz, M. Griebel, Sparse grids. Acta Numer. **13**, 147–269 (2004)
16. K. Chaloner, I. Verdinelli, Bayesian experimental design: a review. Stat. Sci. **10**(3), 273–304, 08 (1995)
17. Y. Chen, J.D. Jakeman, C. Gittelson, D. Xiu, Local polynomial chaos expansion for linear differential equations with high dimensional random inputs. SIAM J. Sci. Comput. **37**(1), A79–A102 (2015)
18. A. Chkifa, A. Cohen, G. Migliorati, F. Nobile, R. Tempone, Discrete least squares polynomial approximation with random evaluations application to parametric and stochastic elliptic PDEs. ESAIM: M2AN **49**(3), 815–837 (2015)
19. P.G. Constantine, Active subspaces: emerging ideas for dimension reduction in parameter studies. SIAM (2015)
20. P.G. Constantine, M.S. Eldred, E.T. Phipps, Sparse pseudospectral approximation method. Comput. Methods Appl. Mech. Eng. **229–232**, 1–12 (2012)
21. P.G. Constantine, E.T. Phipps, T.M. Wildey, Efficient uncertainty propagation for network multiphysics systems. Int. J. Numer. Methods Eng. **99**(3), 183–202 (2014)
22. T. Cui, J. Martin, Y.M. Marzouk, A. Solonen, A. Spantini, Likelihood-informed dimension reduction for nonlinear inverse problems. Inverse Probl. **30**(11), 114015 (2014)
23. R.I. Cukier, H.B. Levine, K.E. Shuler, Nonlinear sensitivity analysis of multi-parameter model systems. J. Comput. Phys. **26**, 1–42 (1978)
24. O. David, J.C. Ascough, W. Lloyd, T.R. Green, K.W. Rojas, G.H. Leavesley, L.R. Ahuja, A software engineering perspective on environmental modeling framework design. Environ. Model. Softw. **39**, 201–213 (2013)
25. L. Devroye, L. Gyorfi, *Nonparametric Density Estimation: The L_1 View* (Wiley, New York, 1985)
26. A. Doostan, H. Owhadi, A non-adapted sparse approximation of PDEs with stochastic inputs. J. Comput. Phys. **230**(8), 3015–3034 (2011)
27. W. Gautschi, *A Survey of Gauss-Christoffel Quadrature Formulae* Birkhäuser Basel, 1981), pp. 72–147
28. G. Geraci, M.S. Eldred, G. Iaccarino, A multifidelity multilevel monte carlo method for uncertainty propagation in aerospace applications, in *19th AIAA Non-Deterministic Approaches Conference* (AIAA SciTech Forum, 2017)
29. T. Gerstner, M. Griebel, Numerical integration using sparse grids. Numer. Algorithms **18** (3–4), 209–232 (1998)
30. T. Gerstner, M. Griebel, Dimension-adaptive tensor-product quadrature. Computing **71**(1), 65–87 (2003)
31. R.G. Ghanem, P.D. Spanos, *Stochastic Finite Elements: A Spectral Approach* (Springer, New York, NY, USA, 1991)
32. M.B. Giles, Multilevel monte carlo methods. Acta Numer. **24**, 259–328 (2015)
33. W.R. Gilks, S. Richardson, D. Spiegelhalter, Markov Chain Monte Carlo in Practice. Chapman & Hall/CRC Interdisciplinary Statistics (Taylor & Francis, 1995)
34. A. Gorodetsky, Y. Marzouk, Mercer kernels and integrated variance experimental design: Connections between gaussian process regression and polynomial approximation. SIAM/ASA J. Uncertain. Quantif. **4**(1), 796–828 (2016)
35. J.B. Gregersen, P.J.A. Gijsbers, S.J.P. Westen, Openmi: open modelling interface. J. Hydroinform. **9**(3), 175–191 (2007)
36. J.H.A. Guillaume, R.J. Hunt, A. Comunian, R.S. Blakers, B. Fu, Methods for exploring uncertainty in groundwater management predictions, in *Integrated Groundwater Management: Concepts, Approaches and Challenges*, ed. by A.J. Jakeman, O. Barreteau, R.J. Hunt, J.D. Rinaudo, A. Ross (Springer International Publishing, Cham, 2016), pp. 711–737

37. J.H.A. Guillaume, M. Arshad, A.J. Jakeman, M. Jalava, M. Kummu, Robust discrimination between uncertain management alternatives by iterative reflection on crossover point scenarios: Principles, design and implementations. Environ. Model. Softw. **83**, 326–343 (2016)
38. E. Haber, Z. Magnant, C. Lucero, L. Tenorio, Numerical methods for a-optimal designs with a sparsity constraint for ill-posed inverse problems. Comput. Optim. Appl. **52**(1), 293–314 (2012)
39. P. Hall, S.J. Sheather, M.C. Jones, J.S. Marron, On optimal data-based bandwidth selection in kernel density estimation. Biometrika **78**(2), 263–269 (1991)
40. J.H. Halton, On the efficiency of certain quasi-random sequences of points in evaluating multi-dimensional integrals. Numer. Math. **2**, 84–90 (1960)
41. J. Hampton, A. Doostan, Compressive sampling of polynomial chaos expansions: convergence analysis and sampling strategies. J. Comput. Phys. **280**, 363–386 (2015)
42. M. Hegland, G. Hooker, S. Roberts, Finite element thin plate splines in density estimation. ANZIAM J. **42**, 712–734 (2009)
43. M. Hohenbichler, R. Rackwitz, First-order concepts in system reliability. Struct. Saf. **1**(3), 177–188 (1982)
44. L. Horesh, E. Haber, L. Tenorio, *Optimal Experimental Design for the Large-Scale Nonlinear Ill-Posed Problem of Impedance Imaging* (Wiley, 2010), pp. 273–290
45. G.M. Hornberger, R.C. Spear, An approach to the preliminary analysis of environmental systems. J. Environ. Manag. **12**, 8–18 (1981)
46. R.W. Hut, N.C. van de Giesen, N. Drost, Comment on most computational hydrology is not reproducible, so is it really science?, in *Let Hydrologists Learn the Latest Computer Science by Working with Research Software Engineers (rses) and not Reinvent the Waterwheel Ourselves*, ed. by C. Hutton et al. (Water Resources Research, 2017)
47. C. Hutton, T. Wagener, J. Freer, D. Han, C. Duffy, B. Arheimer, Most computational hydrology is not reproducible, so is it really science? Water Resour. Res. **52**(10), 7548–7555 (2016)
48. A.J. Jakeman, R.A. Letcher, Integrated assessment and modelling: features, principles and examples for catchment management. Environ. Model. Softw. **18**(6), 491 – 501, 2003. Applying Computer Research to Environmental Problems
49. A.J. Jakeman, R.A. Letcher, J.P. Norton, Ten iterative steps in development and evaluation of environmental models. Environ. Model. Softw. **21**(5), 602–614 (2006)
50. A.J. Jakeman, O. Barreteau, R.J. Hunt, J.D. Rinaudo, A. Ross (eds.), *Integrated Groundwater Management: Concepts, Approaches and Challenges* (Springer International Publishing, 2016)
51. J.D. Jakeman, M. Eldred, D. Xiu, Numerical approach for quantification of epistemic uncertainty. J. Comput. Phys. **229**(12), 4648–4663 (2010)
52. J.D. Jakeman, S.G. Roberts, Local and dimension adaptive stochastic collocation for uncertainty quantification, in *Sparse Grids and Applications*, vol. 88, Lecture Notes in Computational Science and Engineering, ed. by J. Garcke, M. Griebel (Springer, Berlin Heidelberg, 2013), pp. 181–203
53. J.D. Jakeman, T. Wildey, Enhancing adaptive sparse grid approximations and improving refinement strategies using adjoint-based a posteriori error estimates. J. Comput. Phys. **280**, 54–71 (2015)
54. M.A. Janssen, The practice of archiving model code of agent-based models. J. Artif. Soc. Soc. Simul. **20**(1), 2 (2017)
55. J.L. Jennifer, J.M. Gilbert, P.G. Constantine, R.M. Maxwell, Reprint of: active subspaces for sensitivity analysis and dimension reduction of an integrated hydrologic model. Comput. Geosci. **90**, 78–89 (2016)
56. M.C. Jones, J.S. Marron, S.J. Sheather, A brief survey of bandwidth selection for density estimation. J. Am. Stat. Assoc. **91**(433), 401–407 (1996)
57. J. Kaipio, E. Somersalo, *Statistical and Computational Inverse Problems* (Springer, 2005)
58. A. Der Kiureghian, H.Z. Lin, S.J. Hwang, Second-order reliability approximations. J. Eng. Mech. **113**(8), 1208–1225 (1987)

59. J.H. Kwakkel, E. Pruyt, Exploratory modeling and analysis, an approach for model-based foresight under deep uncertainty. Technol. Forecast. Soci. Change **80**(3), 419 – 431, 2013. Future-Oriented Technology Analysis
60. R.J. Lempert, A new decision sciences for complex systems. Proc. Natl. Acad. Sci. **99**(suppl 3), 7309–7313 (2002)
61. R.J. Lempert, D.G. Groves, S.W. Popper, S.C. Bankes, A general, analytic method for generating robust strategies and narrative scenarios. Manag. Sci. **52**(4), 514–528 (2006)
62. R.A. Kelly (Letcher), A.J. Jakeman, O. Barreteau, M.E. Borsuk, S. ElSawah, S.H. HAmilton, H.J. Henriksen, S. Kuikka, H.R. Maier, A.E. Rizzoli, H. van Delden, A.A. Voinov, Selecting among five common modelling approaches for integrated environmental assessment and management. Environ. Model. Softw. **47**, 159–181 (2013)
63. J. Li, J. Li, D. Xiu, An efficient surrogate-based method for computing rare failure probability. J. Comput. Phys. **230**(24), 8683–8697 (2011)
64. Q. Liao, K. Willcox, A domain decomposition approach for uncertainty analysis. SIAM J. Sci. Comput. **37**(1), A103–A133 (2015)
65. H. Liu, J.D. Lafferty, L.A. Wasserman, Sparse nonparametric density estimation in high dimensions using the rodeo, in *AISTATS* (2007), pp. 283–290
66. J.C. Mattingly, N.S. Pillai A.M. Stuart, Diffusion limits of the random walk metropolis algorithm in high dimensions. Ann. Appl. Probab. **22**(3), 881–930 (2012). 06
67. M.D. Morris, Factorial sampling plans for preliminary computational experiments. Technometrics **33**(2), 161–174 (1991)
68. A. Narayan, C. Gittelson, D. Xiu, A stochastic collocation algorithm with multifidelity models. SIAM J. Sci. Comput. **36**(2), A495–A521 (2014)
69. A. Narayan, J.D. Jakeman, T. Zhou, A Christoffel function weighted least squares algorithm for collocation approximations. Math. Comput. **86**, 1913–1947 (2017)
70. L.W.-T. Ng, M. Eldred, Multifidelity uncertainty quantification using non-intrusive polynomial chaos and stochastic collocation, in *53rd AIAA/ASME/ASCE/AHS/ASC Structures, Structural Dynamics and Materials Conference 20th AIAA/ASME/AHS Adaptive Structures Conference 14th AIAA* (2012), p. 1852
71. F. Nobile, R. Tempone, C.G. Webster, A sparse grid stochastic collocation method for partial differential equations with random input data. SIAM J. Numer. Anal. **46**(5), 2309–2345 (2008)
72. S.D. Peckham, E.W.H. Hutton, B. Norris, A component-based approach to integrated modeling in the geosciences. Comput. Geosci. **53**, 3–12 (2013)
73. B. Peherstorfer, K. Willcox, M. Gunzburger, Optimal model management for multifidelity monte carlo estimation. SIAM J. Sci. Comput. (2016). to appear
74. D. Pflüger, B. Peherstorfer, H.-J. Bungartz, Spatially adaptive sparse grids for high-dimensional data-driven problems. J. Complex. **26**(5), 508–522 (2010)
75. C.E. Rasmussen, Gaussian processes to speed up hybrid monte carlo for expensive bayesian integrals, in *Bayesian Statistics*, ed. by J.M Bernardo, A.P. Dawid, J.O Berger, M. West, D. Heckerman, M.J. Bayarri, F.M.A. Smith, vol. 7 (Oxford University Press, 2003), pp. 651–659
76. C.E. Rasmussen, C. Williams, *Gaussian Processes for Machine Learning* (MIT Press, 2006)
77. J.C. Refsgaard, J.P. van der Sluijs, A.L. Hjberg, P.A. Vanrolleghem, Uncertainty in the environmental modelling process a framework and guidance. Environ. Model. Softw. **22**(11), 1543–1556 (2007)
78. G.O. Roberts, J.S. Rosenthal, Optimal scaling for various metropolis-hastings algorithms. Stat. Sci. **16**(4), 351–367 (2001). 11
79. J.O. Royset, R.J.-B. Wets, Fusion of hard and soft information in nonparametric density estimation. Eur. J. Oper. Res. **247**(2), 532–547 (2015)
80. A. Saltelli, R. Bolado, An alternative way to compute Fourier amplitude sensitivity test (fast). Comput. Stat. Data Anal. **26**(4), 445–460 (1998)
81. A. Saltelli, K. Chan, E. Scott, *Sensitivity Analysis* (Wiley, New York, 2004)
82. S. Sankararaman, S. Mahadevan, Likelihood-based approach to multidisciplinary analysis under uncertainty. J. Mech. Des. **134**(3) (2012)
83. D.W. Scott, *Multivariate density estimation: theory, practice, and visualization* (Wiley, 2015)

84. S.A. Smolyak, Quadrature and interpolation formulas for tensor products of certain classes of functions. Soviet Math. Dokl. **4**, 240–243 (1963)
85. I.M. Sobol, Sensitivity estimates for nonlinear mathematical models. Math. Model. Comput. Exp. **1**(4), 407–414 (1993)
86. M. Sobol', B.V. Shukhman, Integration with quasi random sequences: numerical experience. Int. J. Mod. Phys. C **6**(2), 263–275 (1995)
87. R.C. Spear, T.M. Grieb, N. Shang, Parameter uncertainty and interaction in complex environmental models. Water Resour. Res. **30**(11), 3159–3169 (1994)
88. R. Srinivasan, *Importance Sampling: Applications in Communications and Detection* (Springer, 2002)
89. A.H. Stroud, *Approximate Calculation of Multiple Integrals* (Prentice-Hall, Englewood Cliffs, N.J., 1971)
90. A.M. Stuart, Inverse problems: a bayesian perspective. Acta Numer. **19**, 451–559 (2010)
91. B. Sudret, Global sensitivity analysis using polynomial chaos expansions. Reliab. Eng. Syst. Saf. **93**(7), 964–979 (2008)
92. D.G. Tarboton, A. Sharma, U. Lall, Disaggregation procedures for stochastic hydrology based on nonparametric density estimation. Water Resour. Res. **34**(1), 107–119 (1998)
93. G.R. Terrell, D.W. Scott, Variable kernel density estimation. Ann. Stat. **20**(3), 1236–1265 (1992)
94. J.R. Thompson, R.A. Tapiam, Nonparametric function estimation, modeling, and simulation. SIAM (1990)
95. J.P. Van Der Sluijs, M. Craye, S. Funtowicz, P. Kloprogge, J. Ravetz, J. Risbey, Combining quantitative and qualitative measures of uncertainty in model-based environmental assessment: The nusap system. Risk Anal. **25**(2), 481–492 (2005)
96. S. Walsh, T. Wildey, J.D. Jakeman, A consistent bayesian formulation for stochastic inverse problems based on push-forward measures. ASCE-ASME J. Risk Uncertain. Eng. Syst. Part B: Mech. Eng. (2016). accepted
97. R.E. Wengert, A simple automatic derivative evaluation program. Commun. ACM **7**(8), 463–464 (1964)
98. D. Xiu, J.S. Hesthaven, High-order collocation methods for differential equations with random inputs. SIAM J. Sci. Comput. **27**(3), 1118–1139 (2005)
99. D. Xiu, G.E. Karniadakis, The Wiener-Askey polynomial chaos for stochastic differential equations. SIAM J. Sci. Comput. **24**(2), 619–644 (2002)
100. D. Xiu, Numerical integration formulas of degree two. Appl. Numer. Math. **58**(10), 1515–1520 (2008)

Index

R. S. Anderssen et al. (eds.), *Agriculture as a Metaphor for Creativity in All Human Endeavors*, Mathematics for Industry 28,
https://doi.org/10.1007/978-981-10-7811-8

MIX
Papier aus verantwortungsvollen Quellen
Paper from responsible sources
FSC® C105338

If you have any concerns about our products,
you can contact us on
ProductSafety@springernature.com

In case Publisher is established outside the EU,
the EU authorized representative is:
Springer Nature Customer Service Center GmbH
Europaplatz 3, 69115 Heidelberg, Germany

Printed by Libri Plureos GmbH
in Hamburg, Germany